Uranus, Neptune, and Pluto and How to Observe Them

For other titles published in this series, go to
www.springer.com/series/53```

Richard W. Schmude, Jr.

Uranus, Neptune, and Pluto and How to Observe Them

Springer

Richard W. Schmude, Jr.
Gordon College
Barnesville, GA
USA
Schmude@gdn.edu

ISBN: 978-0-387-76601-0   e-ISBN: 978-0-387-76602-7
DOI: 10.1007/978-0-387-76602-7

Library of Congress Control Number: 2008933217

Printed on acid-free paper

springer.com

This book is dedicated to the many people who have helped me along the way: First, to my father and mother, Richard and Winifred Schmude, who first showed me the stars and answered my many science questions; next, to the many fine teachers and professors that I have had along the way; also, to all of the fine people at Optec, Inc., who make a great line of photometers without which I would not have been able to carry out much outer solar system research; and finally, to the many friends who have encouraged me along the way, including Jim Fox and Jerry Sherlin of the Astronomical League, Donald Parker, John Westfall, Ken Poshedly, Richard Jakiel, and Walter Haas of the Association of Lunar and Planetary Observers, Richard McKim and John Rogers of the British Astronomical Association, and Kim Hay of the Royal Astronomical Society of Canada.

Contents

Author's Note

I became interested in astronomy initially when I saw what appeared to be a countless number of stars from my parent's home in Cabin John, Maryland. I was no older than six when I had this life-changing view of the night sky. I purchased my first telescope when I was 15 years old; it had an aperture of 1.5 cm and a magnification of 8X. I enjoyed looking at the Moon and at other distant objects through this telescope. I was also able to let others look through this small wonder, including a couple of siblings and at least one neighbor.

Even as a young boy, I had been fascinated with all of the planets. My first view of Uranus was from my parent's home near Tomball, Texas, on March 20, 1987. In 1989, I began estimating the brightness of Uranus with binoculars, and, a year later, John Westfall asked me to be the outer planets coordinator for the Association of Lunar and Planetary Observers (ALPO). With this appointment, I purchased an SSP-3 solid-state photometer from Optec Inc. and began carrying out brightness measurements of Uranus and Neptune.

Our knowledge of the outer Solar System has increased tremendously in the last 40 years. As a boy, I would often look at astronomy textbooks to learn about the planets. On many occasions, I noticed question marks next to Uranus, Neptune, and Pluto in tables summarizing planetary data. The question marks were there because we did not know much about these objects in the late 1960s. Later, in college, I remember that there was just one page devoted to Uranus, Neptune, and Pluto in my college astronomy textbook; this was in 1981. Once again, we did not know much about these outer worlds and, hence, there was little to write about. With the Voyager 2 flybys of Uranus (1986) and Neptune (1989), we have learned a great deal more about them. The Hubble Space Telescope, electronic cameras, and advanced computer technology have also given us more information about Pluto. In the early twenty-first century, humankind has gained enough information about these distant planets to justify the writing and publication of this book.

This book is broken down into two major sections. The first section summarizes our current knowledge of Uranus (Chapter 1), Neptune (Chapter 2), and Pluto (Chapter 3). The second section describes observing projects that one can carry out with small telescopes and binoculars (Chapter 4), medium-sized telescopes (Chapter 5), and large telescopes (Chapter 6). Finally, an appendix, a bibliography, and an index are included.

Two organizations that are engaged in serious studies of the remote planets are the British Astronomical Association (BAA) and the Association of Lunar and Planetary Observers (ALPO). The current remote planets' coordinator of the BAA is Roger Dymock, who can be reached at: roger.dymock@ntlworld.com, and the current remote planets coordinator of the ALPO is myself, Richard W. Schmude, Jr. I can be reached at: Schmude@gdn.edu. If one makes observations of one of these distant worlds, please let one or both of us know about it. Many thanks!

About the Author

Dr. Richard W. Schmude, Jr., was born in Washington, D.C., and attended public school in Cabin John, Maryland; Los Angeles, California; and Houston, Texas. He graduated from Texas A & M University with a Bachelor's degree in Chemistry, and a few years later, another Bachelor's degree in Physics. Later, he obtained a Master's degree in Physical Chemistry and a Ph.D. in Physical Chemistry. He worked at Los Alamos National Laboratory as a graduate research assistant from 1990 to 1992. While at Los Alamos, he purchased his first photometer and began measuring the brightness of Uranus and Neptune.

In 1990, Richard was appointed Coordinator of the Remote Planets section of the Association of Lunar and Planetary Observers. He began teaching at Gordon College in Barnesville, Georgia, in 1994 and has taught chemistry, physics, astronomy, and physical science there ever since. He has published over 100 scientific papers in many different journals and has given many talks and workshops to community organizations.

The Uranus System

Introduction

Since its discovery in 1781, Uranus has been an object of mystery. It lies 2,900 million kilometers (1,800 million miles) from our Sun. This great distance of the seventh planet, along with its long seasons (about 21 years each), are two reasons why Uranus has been slow to reveal its secrets. Thanks to dedicated scientists and engineers, along with the availability of modern equipment, the growth of our knowledge of Uranus has accelerated since the mid-1980 s. Figure 1.1 shows a series of images of Uranus and its rings made by operators of the modern Keck telescope equipped with advanced imaging technology. Table 1.1 lists a few characteristics of this planet.

Figure 1.2 show the number of reports dedicated to Uranus, Neptune, and Pluto appearing in a journal on the Solar System. The increase in knowledge is due primarily to three factors: (a) Voyager 2, (b) the development of modern astronomical telescopes and equipment and (c) a sharp improvement in computer technology.

In this chapter, we will discuss the atmosphere of Uranus followed by discussions of its bulk composition and interior. This will be followed by a discussion of Uranus's magnetic environment, rings, and moons.

Chapters 2 and 3 present a summary of our current knowledge of the Neptune and Pluto systems, respectively.

Atmosphere

Earth and Uranus have dynamic atmospheres. Both planets obtain most of their heat from the Sun and undergo a cycle of seasons. In this section, we will discuss an altitude reference point for Uranus, followed by a discussion of the gases, clouds, and winds in the Uranian atmosphere. Before these traits are discussed, though, an altitude reference point must be established for Uranus.

All altitudes on Earth are given with respect to sea level; however, Uranus does not have a visible ocean. Hence, an alternative way of describing altitudes must be established. The general convention for Uranus and Neptune is to describe the altitude in terms of the local atmospheric pressure; for example, a cloud at the "2.0 bar level" will be at an area where the pressure is 2.0 bar. One "bar" of pressure

R.W. Schmude, Jr., *Uranus, Neptune, and Pluto and How to Observe Them*,
DOI: 10.1007/978-0-387-76602-7_1, © Springer Science+Business Media, LLC 2008

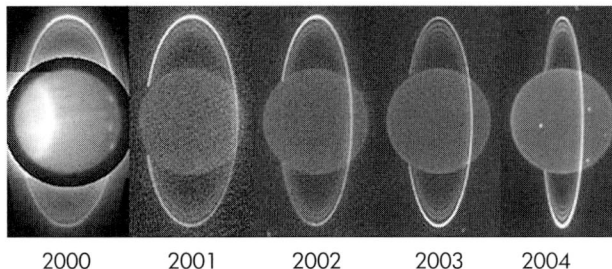

2000	2001	2002	2003	2004

Figure 1.1. Five near-infrared images of Uranus and its rings made with Earth-based telescopes. The rings are easy to see because the methane in Uranus's atmosphere absorbs most of the near-infrared light hitting Uranus causing it to be very dark. The rings are closing up due to the changing orientation of Uranus as seen from Earth. (Credit: Imke de Pater, Heidi Hammel and Sarah Gibbard.)

Table 1.1. Characteristics of Uranus

Characteristic	Value
Equatorial diameter	25,559 km (1 bar level)
Polar diameter	24,974 km (1 bar level)
Surface area	8.08×10^9 km^2 (1 bar level)
Mass	8.68×10^{25} kg
Density	1.27 g/cm^3
Period of rotation	17.24 hours
Period of revolution around the Sun	84.0 years
Inclination	82.2°
Average distance from the Sun	19.19 au
Average distance from Earth at opposition	18.19 au
Orbital inclination	0.8°
Orbital eccentricity	0.05
Ellipticity (or polar flattening)	0.0229
Magnetic field strength at 1 bar level	0.1 to 1.0 Gauss

is about what we experience at sea level here on Earth. For the purpose of defining the diameter of Uranus, an atmospheric pressure of 1.0 bar is selected. Therefore, both diameters in Table 1.1 correspond to an atmospheric pressure of 1.0 bar. The 2.0 bar level is at a lower altitude than the 1.0 bar level. Throughout this book, altitudes are either given with respect to the 1.0 bar level or in terms of the local atmospheric pressure.

Some of the terms used for our atmosphere can also be used for Uranus; however, one must be careful. As an example, Uranus does not have a well-defined mesosphere, or if it does, we have not been able to accurately measure its temperature and density.

The atmosphere of Uranus can be divided into three regions based on altitude, namely, the upper, middle, and lower atmosphere. Each of these regions is discussed below.

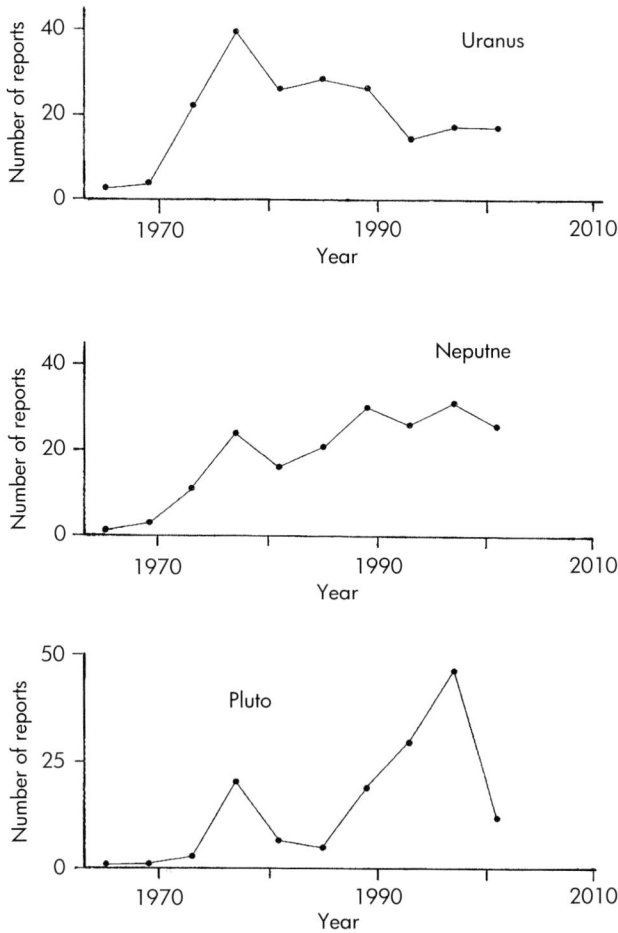

Figure 1.2. Three graphs showing the number of reports that discuss Uranus, Neptune and Pluto appearing in the professional Journal *Icarus* from 1963 to 2004. Each point shows the number of reports published in three-year increments starting with 1963 to 1965. (Credit: Richard W. Schmude, Jr.)

Upper Atmosphere

The upper atmosphere lies above the homopause, which is near the 0.02 mbar (millibar) level. Above the homopause, the percentage of heavier components, such as helium and methane, drop off more quickly with increasing altitude than lighter components, such as atomic hydrogen.

The upper atmosphere contains two types of matter – plasma and gas – and, at some locations; each component exists in the same area. The gaseous component is made up of neutral species. This component contains the thermosphere and exosphere. The ionosphere contains electrons and positively charged ions such as H^+. Figure 1.3 shows the different parts of the upper atmosphere.

Plasma consists of subatomic particles and ions. Its charge is usually balanced, which means that it has an equal amount of positively and negatively charged

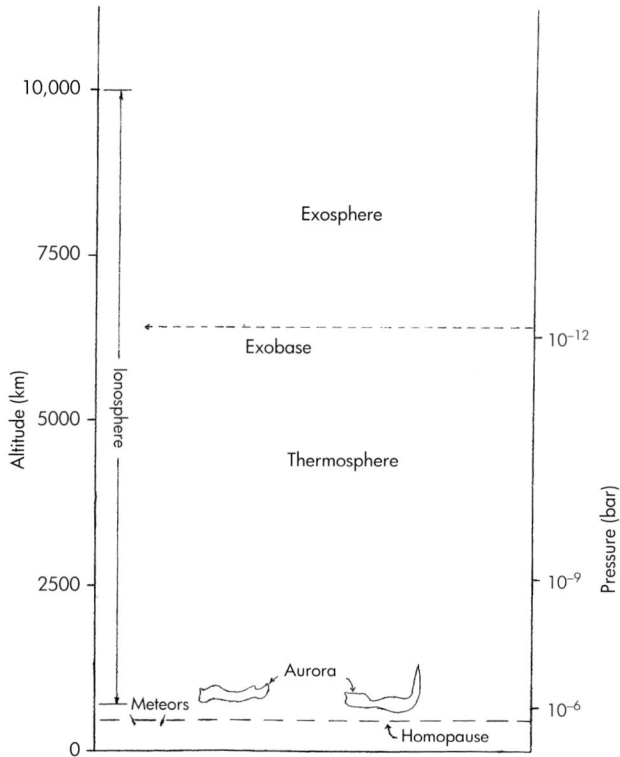

Figure 1.3. A diagram of Uranus's upper atmosphere. The altitude is with respect to the 1.0 bar level. Different levels of pressure are shown at the right. (Credit: Richard W. Schmude, Jr.)

particles. Protons have a positive charge and electrons have a negative charge. Atoms or molecules that have more electrons than protons are called anions, whereas those with more protons than electrons are called cations. Cations and anions are both called ions and can make up a large fraction of some plasmas.

Most of our information of Uranus's ionosphere comes from Voyager 2. This probe passed Uranus in 1986, when the Sun was near a minimum in its sunspot cycle. We know that Earth's ionosphere thins out near solar minimum, and there is a chance that this may have occurred for Uranus in 1986.

Uranus's ionosphere lies at altitudes of 600– $\sim$10,000 km. This altitude range contains both neutral species and charged particles. The ionosphere refers only to the charged particles, such as electrons and protons. Its electron density ranges from $\sim$100 to over 10,000 electrons/cm^3. The peak electron density is at an altitude of 1,500 km; for a comparison, the gas density at this altitude is around 10^{12} molecules/cm^3. There are over 1 million neutral atoms for every ion or electron at this altitude. Despite this, the charged particles may be the dominant heat source at altitudes of 600–10,000 km.

The neutral gas in the upper atmosphere can be broken down into three layers: the bottom layer, which has pressures of between about 2×10^{-5} and 1×10^{-6} bar, then the thermosphere, which has approximate pressures of between about 10^{-6} and 10^{-12} bar, and the exosphere, which has pressures below 10^{-12} bar. Each of these layers is described below, starting with the bottom layer. See Figure 1.3.

Table 1.2. A few photochemical and ionic reactions that are believed to take place in the upper and middle atmospheres of Uranus; the symbol for an electron is e^-

Upper Atmosphere

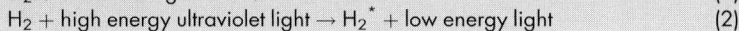

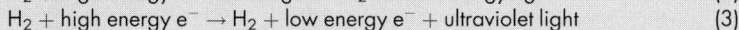

H_2 + ultraviolet light $\rightarrow$ H + H	(1)
H_2 + high energy ultraviolet light $\rightarrow$ H_2^* + low energy light	(2)
H_2 + high energy e^- $\rightarrow$ H_2 + low energy e^- + ultraviolet light	(3)

*Hydrogen is in a high energy (or excited) state.

Ionosphere

$H_2 + e^- \rightarrow H^+ + H + 2\,e^-$	(4)
$He + e^- \rightarrow He^+ + 2\,e^-$	(5)

Middle atmosphere (or stratosphere)

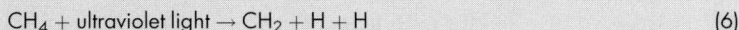

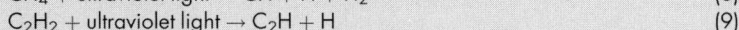

CH_4 + ultraviolet light $\rightarrow$ CH_2 + H + H	(6)
CH_4 + ultraviolet light $\rightarrow$ CH_2 + H_2	(7)
CH_4 + ultraviolet light $\rightarrow$ CH + H + H_2	(8)
C_2H_2 + ultraviolet light $\rightarrow$ C_2H + H	(9)
$CH_4 + CH \rightarrow C_2H_4 + H$	(10)

The bottom layer includes the homopause and the transition between the middle and upper atmosphere. Several important processes occur in this layer. Small pieces of space dust falling into Uranus's atmosphere undoubtedly burn up there and would appear as meteors. The aurora also occurs in this layer. A large amount of ultraviolet light is emitted there as well. Astronomers believe that the two sources of this ultraviolet emission are fluorescence (reaction 2 in Table 1.2) and high-energy electrons colliding with neutral species, like hydrogen (H_2), creating ultraviolet light (reaction 3 in Table 1.2). Fluorescence is a process whereby material absorbs light with a high energy and then emits lower energy light.

Astronomers can probe the bottom layer of the upper atmosphere using Earth-based stellar occultation data. Occultation data show that there is a small temperature increase at around the 3-μbar (microbar) level. One group of astronomers suggests that this may be due to material falling from the rings, which is heated by sunlight.

The thermosphere has several distinct characteristics. It starts at an altitude of 600 km and extends upwards for a few thousand kilometers. This region gets much of its heat from the ionosphere. At an altitude of 1,500 km, the temperature is approximately 500–600 K and rises with increasing altitude. Another characteristic of the thermosphere is that its composition changes with altitude. In one study, the thermosphere is reported to have over 99% molecular hydrogen (H_2), with just traces of helium and atomic hydrogen (H) at an altitude of 1,500 km. Furthermore, the composition is predicted to change to 90% H_2 and 10% H at an altitude of 5,000 km and at altitudes above 10,000 km, H is predicted to be more abundant than H_2. The reason for this is that H has a lower molar mass (1 g/mole) than H_2 (2 g/mole). Gases with high molar masses thin out quicker at increasing altitudes in the thermosphere than those with low molar masses. The thermosphere emits some ultraviolet light.

Above the thermosphere lies the exosphere, which has characteristics different from the lower atmospheric layers. One group reports that this layer scatters ultraviolet light out to at least 70,000 km above the 1.0 bar level. Ring particles experience drag forces from the exosphere. The outer portion of the exosphere is often called the corona.

The outer part of the exosphere is far from Uranus. Atoms in this area that are moving from Uranus are unlikely to collide with other atoms and thus may escape from the planet. Atoms below the exosphere, which are moving from Uranus, however, are in a dense enough gas that they will probably collide with other atoms. As a result, gases below the exosphere are much less likely to escape than those in the exosphere. The bottom border of the exosphere is the exobase; for Uranus, this is about 6,400 km above the 1.0 bar level.

Middle Atmosphere (Stratosphere)

The middle atmosphere is Uranus's stratosphere. This is a region that lies above the 0.1 bar level, and its temperature rises with increasing altitude. The upper boundary of the stratosphere is difficult to pinpoint due to the problem of identifying Uranus's mesosphere; let's call it the 0.020 mbar level, which is near the homopause. Unlike the upper atmosphere, the chemical composition of the stratosphere does not change much with altitude. This is because gas mixing is more important than at higher altitudes.

The stratosphere is often transparent to visible light for two reasons. First, it often lacks opaque clouds, and second, it is made up mostly of hydrogen and helium; both of these gases are transparent to this light at low pressures. The composition of the stratosphere is listed in Table 1.3. One reason why there are no gases in significant quantities in the stratosphere besides hydrogen, helium, and methane is because of its low temperature. Most compounds (such as ammonia, water, and carbon dioxide) that are pushed upwards would condense at the low temperatures ($\sim$52 K) at the bottom of the stratosphere. Essentially, this area acts like a cold trap. Methane's vapor pressure at 52 K is $\sim$10^{-5} bar, which is high enough for some of that material to pass through the coldest part of the stratosphere as a gas.

Table 1.3. Composition of Uranus's middle atmosphere (or stratosphere)

Component	Percentage by Volume	Mass Fraction
Molecular Hydrogen (H_2)	85%	74%
Deuterated Hydrogen (HD)	Trace	Negligible
H_3^+	Trace	Negligible
Helium (He)	15%	26%
Methane (CH_4)	0.01 to 0.001%	<0.1%
Methyl radical (CH_3)	Trace	Negligible
Methane with deuterium (CH_3D)	Trace	Negligible
Acetylene (C_2H_2)	Trace	Negligible
Ethene (C_2H_4)	Trace	Negligible
Ethane (C_2H_6)	Trace	Negligible
CH_3C_2H or C_3H_4	Trace	Negligible
Diacetylene (C_4H_2)	Trace	Negligible
Carbon Monoxide (CO)	Trace	Negligible
Carbon Dioxide (CO_2)	Trace	Negligible
Water (H_2O)	Trace	Negligible

Several photochemical reactions occur in the middle atmosphere, and a few of these are listed in Table 1.2. Essentially, the small amount of methane and other hydrocarbons react with sunlight to produce a variety of hydrocarbon molecules. Many of these are present in trace quantities and are listed in Table 1.3. These gases condense at the low stratospheric temperatures and create haze. The haze particles are believed to form when material condenses onto microscopic solid particles. Two areas at 0.5 mbar and 13 mbar have higher temperatures than the surrounding altitudes. The condensation temperatures of diacetylene (C_4H_2) and ethane (C_2H_6) are near the 0.5 and 13 mbar levels. These two gases may condense, forming haze layers, which then absorb extra sunlight, thus causing higher temperatures. Additional chemical reactions may take place also on the haze particles. In fact, the haze particles probably contain some hydrocarbon molecules such as C_4H_2 and C_6H_2, which have at least four carbon atoms.

A thorough analysis of Voyager 2 images, especially at high solar phase angles, has given us information about the haze particles near Uranus's south polar region at that time (1986). One group reports that their calculations are consistent with the haze having a low optical depth for visible light, which means that it transmits almost all of this type of light. A second group reports that the haze particles near the 1.0 mbar level have diameters of $\sim$0.01 μm, but, as they fall, they merge with one another, growing to about ten times their original diameters. It takes haze particles several years to fall into the warmer parts of the troposphere.

Methane is recycled in Uranus's atmosphere through the methane cycle. See Figure 1.4. Essentially, methane is destroyed by ultraviolet light and fast-moving

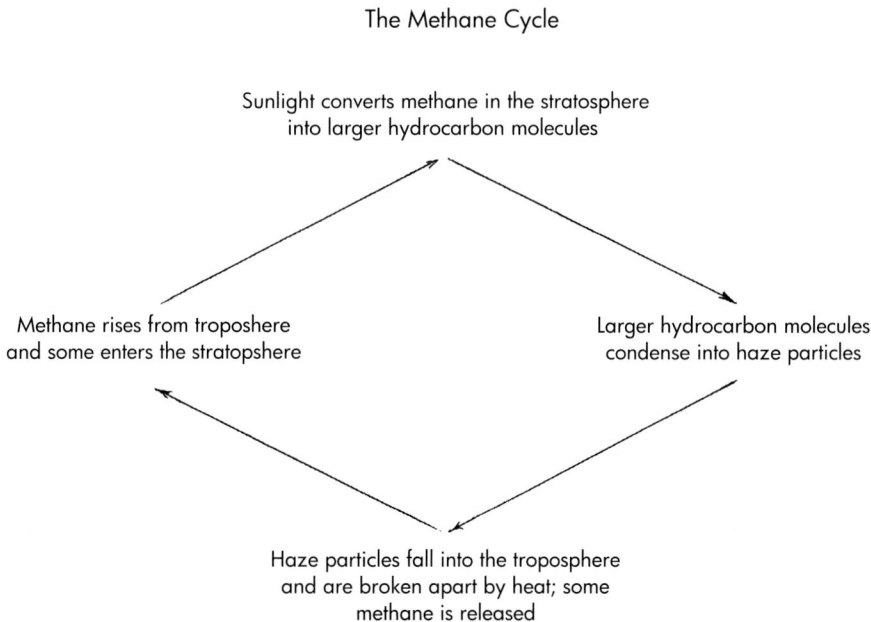

The Methane Cycle

Sunlight converts methane in the stratosphere
into larger hydrocarbon molecules

Methane rises from troposhere
and some enters the stratopshere

Larger hydrocarbon molecules
condense into haze particles

Haze particles fall into the troposphere
and are broken apart by heat; some
methane is released

Figure 1.4. The methane cycle on Uranus. Methane in the stratosphere is converted into larger hydrocarbons, which then condense into haze. The haze falls into the troposphere where the temperatures rise to the point where the larger hydrocarbons are broken down into methane. Some of the methane rises to the stratosphere where the process is repeated. (Credit: Richard W. Schmude, Jr.)

charged particles. It is replenished, though, when haze particles fall into the troposphere. When this occurs, the particles are broken apart by heat and methane is released. Some of this methane works its way back up into the stratosphere.

Lower Atmosphere

The lower atmosphere includes the troposphere, which lies below the tropopause – the coldest part of the atmosphere. Figure 1.5 shows the different cloud and haze layers that may be present in the middle and lower atmosphere. When one looks at Uranus through a telescope, he or she sees the troposphere. The atmosphere down to ~1 bar is usually transparent in visible light. The troposphere at around the 1.0 bar level contains ~83% hydrogen, ~15% helium and 1–2% methane (by volume). The methane concentration rises with increasing depth below the 1.0 bar level. Figure 1.6 shows a nearly true color image of Uranus when Voyager 2 was on the dark side of the planet.

Uranus did not produce much internal heat in 1986, and, as a result, there probably was not much convection in the lower atmosphere. This lack of convection may have led to the absence of ammonia and other compounds at the 1–3 bar level. This, in turn, may have affected the types of hazes and clouds that developed there.

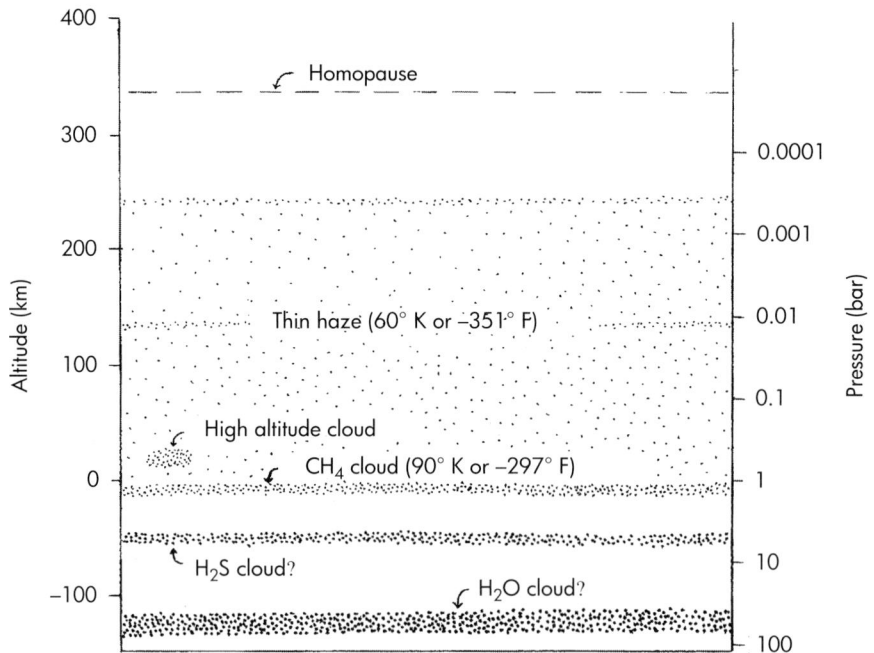

Figure 1.5. A diagram showing how Uranus's middle and lower atmosphere may appear. The altitudes are with respect to the 1.0 bar level. Different levels of pressure are shown at the right. (Credit: Richard W. Schmude, Jr.)

Figure 1.6. A Voyager 2 color image of Uranus and its dark side: This image was taken through three different color filters and was recombined to produce this image. (Credit: Courtesy NASA/JPL-Caltech.)

Chemical reactions may be another reason for the lack of different compounds at the 1–3 bar level. One gas, carbon monoxide (CO), if present, would be converted quickly into methane through the reaction:

$$CO(g) + 3H_2(g) \longrightarrow CH_4(g) + H_2O(g) \tag{1.1.}$$

The (g) represents the gas phase. The low amount of ammonia (NH_3) may be due to the reaction:

$$NH_3(g) + H_2S(g) \longrightarrow NH_4SH(s) \tag{1.2}$$

where the (s) in reaction 1.2 means the solid phase.

Clouds

Before we discuss clouds on Uranus, let's review how clouds form and how one estimates cloud altitudes.

What is a cloud? On Earth, most clouds are made up of millions of either microscopic liquid water droplets or microscopic ice particles. Microscopic particles undoubtedly make up the clouds on the other planets as well.

Clouds on Earth form when the relative humidity exceeds 100% and when microscopic solid particles are present where gases can condense. The relative humidity (RH) is defined as:

$$RH = (AC/MC) \times 100\% \tag{1.3}$$

where "AC" is the actual amount of gaseous water in the air and "MC" is the maximum amount of gaseous water that the air can hold. Hot air can hold more gaseous water than cold air, so MC drops as the temperature drops. Oftentimes, air will contain lots of gaseous water and, when it cools, it causes the RH to reach 100%; at this point, some of the gaseous water condenses into microscopic droplets, forming clouds.

The situation is similar for Uranus, except that one is dealing with methane instead of water in the upper troposphere. The atmosphere can hold only a specific amount of gaseous methane. When it cannot hold any more, the methane either condenses into a cloud if condensation nuclei are present, or becomes a super-cooled vapor if no nuclei are present.

Astronomers use different wavelengths to estimate cloud altitudes. Let me explain how this works. Essentially, light is absorbed and scattered by the gases in Uranus's atmosphere. At pressures greater than $\sim$0.1 bar, hydrogen and helium begin absorbing visible and near-infrared light. Methane also absorbs light. These gases absorb some wavelengths more than others and, as a result, different wavelengths of light sample different depths in Uranus's atmosphere. One example is that infrared light having a wavelength of 1.665 μm reaches the $\sim$1.0 bar level, whereas 2.12 μm light only reaches down to the $\sim$0.3 bar level. Therefore, if a bright cloud is imaged with 1.665 μm light, but not 2.12 μm light, it probably lies between the 0.3 bar and 1.0 bar levels. Essentially, this cloud lies below the 0.3 bar level because it did not show up in 2.12 μm light; however, it lies above the 1.0 bar level because it did show up in 1.665 μm light. In many cases, clouds at a specific altitude will affect one type of light but not another type, due to absorption by the overlying atmosphere. Therefore, images made with different types of light probe different depths in Uranus's atmosphere.

We can learn about the altitudes of the different clouds from multi-wavelength studies. When a high-altitude cloud is imaged, what is really happening is that either a low-lying cloud has developed a high-altitude top where the light is reflected, or light is reflected by a high-altitude cloud. In either case, more light is reflected by the cloud than the surrounding areas and, as a result, it is bright. In Figure 1.7, the low-altitude cloud reflects little visible light because the overlying gas layer absorbs it. The high-altitude cloud, however, reflects more light.

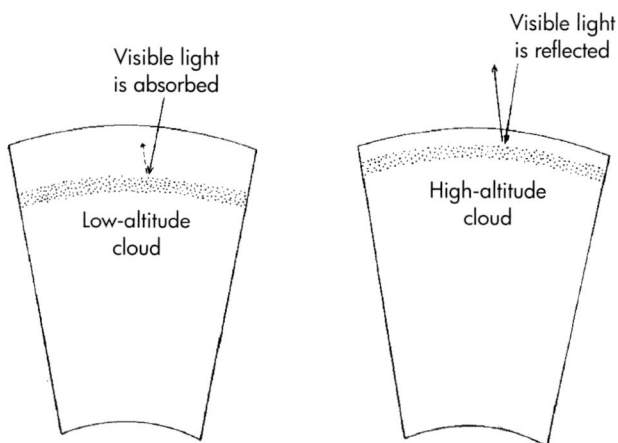

Figure 1.7. High-altitude clouds (right) reflect visible light before it is absorbed by the deeper layers of Uranus's atmosphere. Low-altitude clouds (left), on the other hand, do not reflect as much visible light because it has to pass through a thicker portion of the atmosphere; hence, more of it is absorbed. As a result, high-altitude clouds are brighter than low-altitude clouds in this example. (Credit: Richard W. Schmude, Jr.)

We are not sure of the cloud layers in Uranus's atmosphere. Either a thin methane cloud layer or scattered methane clouds may lie near the one to two bar level. This cloud may change with Uranus's seasons. Temperatures at this level are close to those at which methane condenses. A second and thicker cloud layer may lie near the four bar level. We can only speculate on the chemical composition of this cloud; it may be composed of hydrogen sulfide (H_2S) or ammonia (NH_3). Many astronomers believe that a third cloud layer is at a depth of several tens of bars. Uranus's microwave and radio emissions are consistent with a deep cloud layer.

One recent study is consistent with a haze extending down to several bar, without a methane cloud layer near the one to two bar level. This study is evidence that we still have a lot to learn about Uranus's atmosphere, but that progress is being made!

Astronomers report that the intensity of radio waves at a wavelength of 3.5 cm underwent a cyclic change between 1966 and 2005. The intensity reached a maximum in the mid-1980 s. One interpretation of this data is that the radio waves originate in Uranus's interior and that the atmospheric opacity to radio waves changes between Uranus's south pole and equator. This change may be due to a difference in the opacity of the deep ~50 bar cloud layer between polar and equatorial latitudes.

On Earth, several changes in our atmosphere occur such as cloud formation, cloud dissipation, lightening and precipitation. Most or all of these same processes occur now on Uranus. The Voyager 2 probe detected bursts of radio waves which are probably lightning bolts. We are not sure which cloud layer is producing the lightning. No lightning flashes from Uranus were imaged by Voyager 2. Recent studies have shown that clouds on that planet spring up and dissipate. Some clouds can change in as little as an hour, while others can last more than a month. We are not sure if rain occurs on Uranus. Small amounts of haze may fall to lower altitudes. The deep layers of Uranus's atmosphere may produce methane rain.

Starting in the 1980 s, scientists began studying the movement of clouds on Uranus. Cloud movement shows the wind speed. Cloud movement faster than the planet's rotation period of 17.24 hours is prograde movement, and the wind speed is positive; otherwise, the movement is retrograde with a negative speed. Clouds that rotate once every 17.24 hours have a speed of 0 m/s. Wind speeds on Uranus change with latitude. Winds near the equator move slower than those at around 40°N or S and this causes clouds to rotate at different rates. See Figure 1.8. This also occurs on Neptune, but it is different from Jupiter and Saturn, where the winds near the equator are almost always faster than those at around 40°N or S. We are not sure why equatorial winds on Uranus are so slow.

The highest wind speed measured for Uranus is 218 m/s (488 miles/hour). This is higher than Jupiter's maximum wind speed but is less than the corresponding speed on Saturn. There is a chance that the winds on Uranus change. One group reports that the winds between 20°N and 40°N accelerated between the late 1990 s and 2003.

Astronomers have used visible and near-infrared light to image Uranus's clouds. We cannot see near-infrared light, but electronic cameras can detect it. Most clouds imaged from Earth-based telescopes appear to be 1,000–2,000 km across. We are not sure whether each cloud is a single feature or is a group of smaller clouds.

Thanks to the quality of the Keck and Hubble Space Telescopes, astronomers are now beginning to understand a few statistics of cloud development on Uranus. Two groups of astronomers imaged several dozen clouds on Uranus in 2003 and

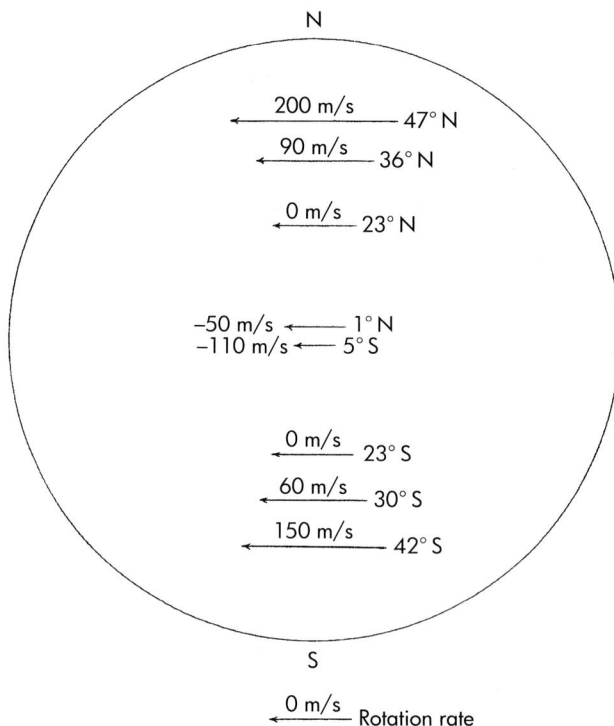

Figure 1.8. Different latitudes of Uranus rotate at different speeds. The large circle is Uranus and the length of the arrow is proportional to the rotation speed. If the wind speed is positive, the wind is prograde; otherwise it is retrograde. Features moving with prograde winds rotate faster than Uranus's rotation rate of 17.24 hours; otherwise they rotate slower. In this figure, the winds at 47°N, 36°N, 30°S and 42°S are prograde whereas those at 1°N and 5°S are retrograde. (Credit: Richard W. Schmude, Jr.)

2004. Most of the clouds were restricted to two areas; 20°S to 50°S (28% of clouds) and 0°N to 50°N (68% of cloud features). The northern hemisphere had more than twice the number of clouds as the southern hemisphere. These clouds were visible in J filter and H filter images. Is this distribution dependent on the season, or is it fixed? Only further observations will answer this question.

One high-altitude cloud developed in July 2004 near 36°S. We know that the cloud had a high altitude because it was bright in 2.12 μm light, which only probes layers above about 0.3 bar. Other bright clouds were not visible in this filter because they were at altitudes below the 0.3 bar level. The discovery of this cloud led one astronomer to state that Uranus has gone from "boring and unchanging" to "interesting and variable". This cloud is interesting because of its high altitude.

Cyclic Changes

There are two cyclic changes of sunlight on Uranus – diurnal and seasonal. Diurnal changes occur as a result of rotation. Diurnal changes are minimal when the poles are nearly facing the Sun, as was the case in 1986. When Uranus is near an equinox point, the situation is different – all areas receive some sunlight during each rotation and, hence, may experience temperature changes. Temperature changes may affect cloud development, so it is important to understand these changes and their magnitude at different atmospheric altitudes.

The changing amounts of sunlight will not affect all parts of Uranus's atmosphere in the same way. Some parts receive lots of sunlight when the Sun lies overhead and, as a result, these areas warm up quickly. Deeper layers, say below the four bar level, will not receive much sunlight even when the Sun is directly overhead; hence, these layers will not warm up at the same rate as the higher altitude areas. There is a good chance that diurnal and seasonal temperature changes are smaller for deep layers compared to ones higher up.

One astronomer at Lowell Observatory carried out a series of measurements in April 1975 and was unable to detect any periodic brightness change exceeding 0.005 magnitudes due to rotation. He used the Stromgren b and y filters. One ALPO member carried out similar studies in August and September 2001 and October 2006 with the broadband V filter. He also detected no diurnal brightness change that exceeded 0.02 magnitudes. These results show that all longitudes of that planet had nearly the same brightness in visible light when they faced the Sun during the times of observation.

A UK amateur astronomer carried out a series of brightness measurements on six nights close to opposition in September 2007. He used filters that were transformed to the Johnson V and cousins Ic system. His results are consistent with a small diurnal brightness changes amounting to a little over one percent in the Ic filter with a period close to that of Uranus's magnetic field. The diurnal brightness change in the V filter, if present, is only a small fraction of one percent.

Does Uranus undergo seasonal changes even though its axis lies close to the ecliptic plane? Yes; in fact, during one 84-year revolution its polar regions receive more sunlight than the equatorial regions. Each of the seasons on Uranus is about 21 years long. The long winters mean that some parts of the atmosphere receive no sunlight for many years, and, as a result, this may affect the atmospheric dynamics. At the time of this writing, we are beginning to see several trends that may be due either to the seasonal cycle of sunlight or to the cycle of changing viewing geometry. Between 1985 and 2007, Uranus's southern hemisphere experienced summer while the northern hemisphere experienced winter. [The International Astronomical Union (IAU) convention is used here.] Starting in late 2007, the southern hemisphere started experiencing fall and the northern hemisphere started experiencing spring; and, for the first time in over 40 years, sunlight fell on Uranus's north pole.

Figure 1.9 shows a recent Hubble Space Telescope image of Uranus and some of its cloud belts. The image was made with three different near-infrared filters, and it shows a bright south polar region. The small white dot is the moon Ariel, and the dark circle is the shadow cast by that moon. Satellite transits will not have much of an effect on the brightness of Uranus.

Peak temperatures may not correspond to high amounts of sunlight on Uranus. One group predicts the highest polar temperatures at the 540 mbar level occur at equinox, which is not the time of maximum solar insulation. They predict that seasonal temperatures lag behind solar insulation by about a quarter of a Uranian year. The temperatures near Uranus's south pole are predicted to rise from 58 K (near spring equinox – 1968) to 60 K (near autumn equinox – 2005) at the 540 mbar level.

Figure 1.9. A false color image of Uranus made from three different near-infrared images. The bright area at the left is the south polar cap. The bright circular spot is the moon Ariel and the dark spot is Ariel's shadow. (Credit: NASA, ESA, Larry Sromovsky, Heidi Hammel and Kathy Rages.)

Long-Term Brightness Changes

During the last half century, Uranus has undergone a cyclic change in brightness that may be related to its seasons. Seasonal changes in either temperature or viewing geometry may affect the brightness of Uranus. In fact, one astronomer published a paper describing long-term brightness changes over 70 years ago. When the polar areas faced the Sun and Earth during solstice, Uranus was usually brighter than when its equator faced us. This is shown in Table 1.4. More recently, people have measured the brightness of Uranus with electronic equipment. One

Table 1.4. Seasonal brightness changes observed from visual magnitude estimates

Year	Seasonal time	V_o (vis)[a]
1882	Equinox	5.7
1901	Solstice	5.4
1923	Equinox	5.7
1946	Solstice	?
1966	Equinox	(5.2)
1985	Solstice	5.6
2006	Near Equinox	5.7

[a] This symbol is for the human eye which has a different sensitivity than the V filter.

group reports that Uranus gradually dimmed near the 1966 and 2007 equinoxes but brightened near the 1985 solstice. The brightness changes from one year to the next were gradual, but, over several years, they were substantial.

Lowell observatory astronomers measured a brightness drop of 0.03 and 0.10 magnitudes for Stromgren b and y filters between 1985 and 2004. Using a filter transformed to the Johnson V system, members of the Association of Lunar and Planetary Observers (ALPO) measured a 0.06 magnitude drop between 1991 and 2006. See Figure 1.10. All measurements in the figure were normalized to average V(1,0) values. This normalization procedure is discussed further in Chapter 5. The Lowell and ALPO results are consistent with Uranus undergoing a brightness change of ~0.1 magnitudes in several visible wavelengths between its 1985 solstice and its equinox in 2007. The corresponding brightness change in the Stromgren b filter is less than half of what it is in the Stromgren y filter. Members of the ALPO used filters transformed to the Johnson B and V system and report that Uranus did not dim as much in the B filter as in the V filter. They have also measured large brightness drops in filters transformed to Johnson red (wavelength = 0.70 μm) and infrared (wavelength = 0.86 μm) system between 1993 and 2006: a change of over 0.2 magnitudes. The B, V, and R filter data are consistent with the planet undergoing a seasonal color change. Uranus was redder near its 1985 solstice than near its equinox in 2006. The photometric data, therefore, show that the planet undergoes both a color change and a brightness change which matches its seasons. Digital images have shown that the Uranian polar regions have a different color and albedo (the fraction of light that a surface reflects) than areas near the equator.

What is causing the long-term brightness change on Uranus in visible light? There may be as many as three factors at work. First, a small amount (~0.025 magnitudes) of this change in visible and near-infrared light is caused by the ellipticity of Uranus. Essentially, we see more of Uranus when one of its polar regions faces us than when the equator faces us. This is described further in Chapter 5. Second, the seasonal development and dissipation of clouds may contribute to brightness changes. Finally, we know that the planet becomes dimmer as a result of the movement away of its brighter south polar region. The south polar region is brighter than the temperate and equatorial regions, perhaps because it has a thicker methane cloud layer. This may contribute also to

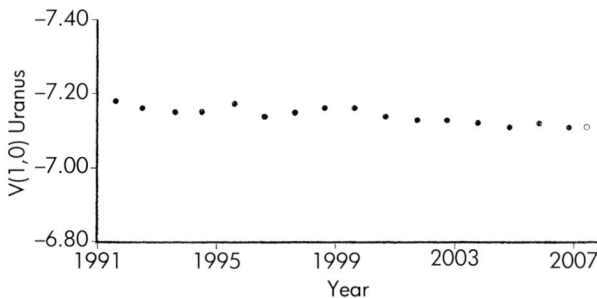

Figure 1.10. The normalized magnitude of Uranus, V(1,0), plotted for different years. All data were collected by members of the Association of Lunar and Planetary Observers (ALPO) using a filter transformed to the Johnson V system. The open circle is one measurement made in 2007 and the filled circles are yearly averages. (Credit: Richard W. Schmude, Jr.)

brightness changes. Perhaps the seasonal development of clouds along with albedo differences between polar and equatorial regions play a role in Uranus's seasonal brightness changes.

One possible scenario for the changes in both visible and radio wavelengths is the two-cloud model shown in Figure 1.11. This model can explain both the seasonal changes in visible light just discussed, and the larger amount of radio waves given off by Uranus when it was at solstice, which was described earlier. In this model, Uranus has two cloud layers, the lower layer at the ~50 bar level blocks radio waves with a wavelength of 3.5 cm coming from the interior, but the upper layer does not affect these waves. An upper cloud or haze layer scatters visible light, but the lower cloud layer does not affect this light. The deep cloud layer is

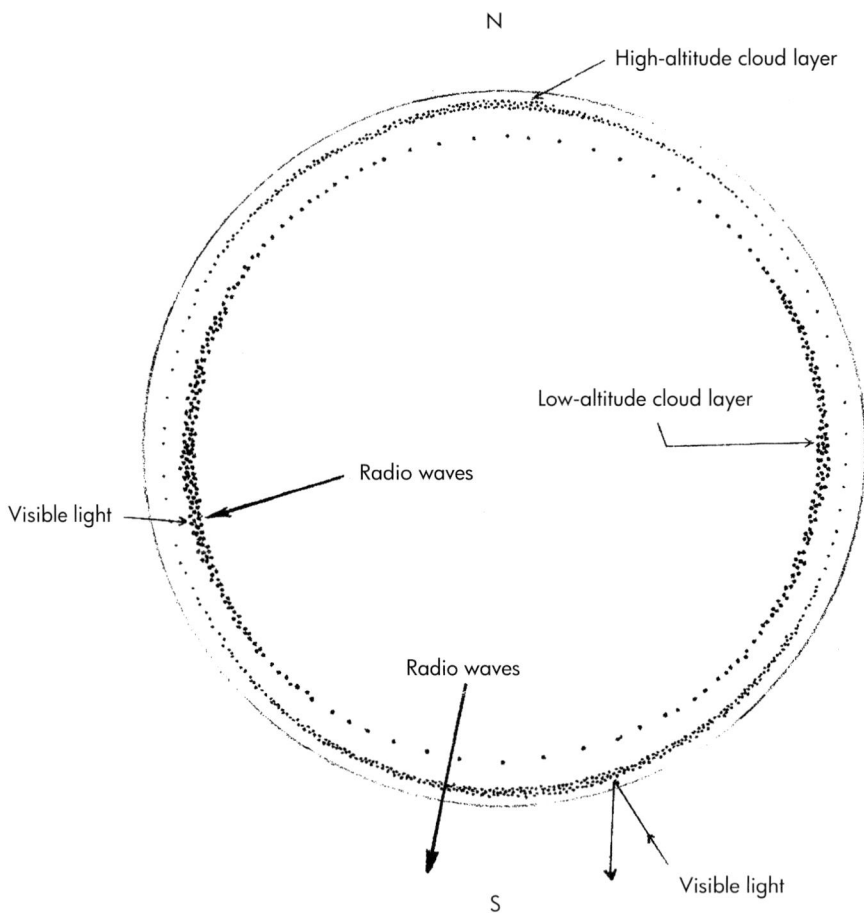

Figure 1.11. A two-cloud layer model of Uranus. In this model, the low altitude cloud layer absorbs radio waves coming from the interior. It is thickest near the equator and is thinnest near the poles. This could explain the low emission when Uranus's equator faced us in 1966 and 2005, and the high emission when that planet's south polar region faced us in 1985. The higher altitude cloud layer reflects visible light, but does not absorb radio waves. If it is highest and thickest near the poles, it could explain the larger amount of visible light reflected by Uranus when its polar region faced the Earth. (Credit: Richard W. Schmude, Jr.)

thickest near the equator and thinnest near the south pole. When the south polar region faced us in the 1980 s, more radio waves came from Uranus because the deep layer was thinner and let more of these waves through, but when the equatorial regions faced us, the radio emissions dropped. Visible light is affected by the upper cloud layer. When the polar region faced us, the upper cloud layer reflected more light; hence, Uranus was brighter. When the equatorial region faced us, the thinner upper cloud reflected less visible light, causing Uranus to be dimmer.

What is the average normalized magnitude of Uranus, and do the northern and southern hemispheres of that planet reflect the same amount of visible light? This question is examined here. The middle point between equinox and solstice occurs when its sub-Earth latitude is ~40°. This occurred in 1956 and 1997, and, as a result, brightness and color measurements centered on these years should yield seasonal averages. The average Vo value between 1954 and 1959 was Vo = 5.568. This value is near the seasonally average magnitude of Uranus's northern hemisphere. The average Vo value of ALPO measurements made between 1991 and 2003 is Vo = 5.57, which is near the seasonally average value for the southern hemisphere. These results show that both hemispheres of Uranus reflect nearly the same amount of visible light. Seasonally averaged photometric constants for Uranus are summarized in Table 1.5.

Astronomers have also measured another photometric constant of Uranus – the solar phase angle coefficient. (The solar phase angle coefficient shows how quickly an object brightens as the solar phase angle drops.) Uranus's average value at solar phase angles of between 0° and 3° is ~0.0017 magnitude/degree for the Stromgren

Table 1.5. Photometric constants for Uranus. These values correspond to seasonal averages

Characteristic	Value
y_o [a]	5.54 ± 0.01
$y(1,0)$	-7.17 ± 0.01
b_o [a]	5.76 ± 0.01
$b(1,0)$	-6.95 ± 0.01
b–y color index	0.22 ± 0.01
Vo [b]	5.57 ± 0.01
$V(1,0)$	-7.15 ± 0.01
B–V [c]	0.53 ± 0.01
V–R [d]	-0.22 ± 0.03
V–I [d]	-1.2 ± 0.1
Bolometric Bond albedo [e]	0.31 ± 0.01
Solar phase angle coefficient (b filter) [f]	0.0027 ± 0.0012 mag./degree
Solar phase angle coefficient (y filter) [f]	0.0017 ± 0.0008 mag./degree
Solar phase angle coefficient, c_V, (V filter) [g]	0.0011 ± 0.0010 mag./degree

[a] Average of 1993 to 2001 (Lowell).
[b] Average of 1991 to 2003 (ALPO) and 1954 to 1959 (Lowell).
[c] Average of 1993 to 2001 (ALPO) and 1954 to 1959 (Lowell).
[d] Average of 1993 to 2001 (ALPO).
[e] From Bergstralh et al (1991) p. 249.
[f] From Lockwood and Thompson (1999); uncertainties are the average standard deviation.
[g] ALPO average for the 1992, 2001, 2002 and 2006 apparitions.

y filter and 0.0027 magnitude/degree for the Stromgren b filter. The average solar phase angle coefficient for the broadband V filter is ~0.0011 magnitude/degree. These values show that Uranus does not have an opposition surge and that it does not brighten much as its solar phase angle drops from 3° to 0°.

Other Characteristics of the Atmosphere

Astronomers have made many fine Uranus images since Voyager 2 that show several albedo features. Here we will discuss a few of them, but be aware that many of these were imaged in near-infrared light. One set of Keck images of Uranus is shown in Figure 1.12. The images were made with J, H, and K filters. The bright polar collar in the J and H filter images is between planetographic latitudes of 42°S and 50°S. One faint dark belt in the J filter images lies between

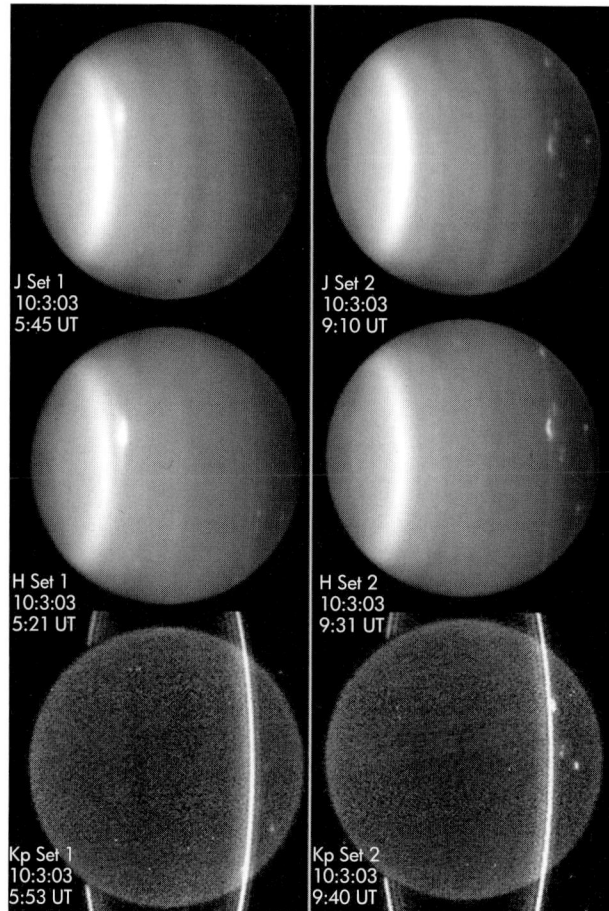

Figure 1.12. Keck images of Uranus made in the J filter (top row), H filter (middle row) and the K filter (bottom row). Images on the left were made between 5:21 and 5:53 UT on Oct. 3, 2003 and those on the right were made a few hours later. In all images the South pole is to the left. (Credit: Heidi Hammel, Imke de Pater, Sarah Gibbard, G. W. Lockwood, and Kathy Rages.)

24°S and 28°S, while a second one lies between 2°S and 8°S. The two dark belts do not show up in K filter images. This is due to the fact that the K filter is sensitive to light at high altitude, whereas the J filter is sensitive to light at lower altitudes. One scientist reports that the light transmitted by the J, H, and K filters is sensitive to the ∼2 bar, ∼1 bar, and ∼0.3 bar levels. The bright polar collar shows up in H filter images but not in K filter images. This collar may lie near the 1.0 bar level. In addition to these three belts, the area north of ∼20°N is darker than other areas in the J and H filter images. Finally, a faint, dark cap is visible above the south polar region. This feature extends down to ∼71°S and is visible in H filter images made in August and October 2003. Scientists analyzing red and infrared Hubble Space Telescope images detected several changes in the south polar region. One of those was that the area south of ∼70°S darkened between 1994 and 2002. The development of a polar haze could explain this change.

One group reported the presence of a faint north polar collar in 2007 that was a bit brighter than the surrounding areas. Only future images will show whether this feature becomes brighter as more sunlight reaches northern latitudes.

One group of astronomers discovered a dark cloud in a red filter image of Uranus using the Hubble Space Telescope. This cloud was near 27°N. It had a north-south dimension of 1,700 km and an east-west dimension of 3,000 km. This feature was about one-fourth the size of similar dark spots on Neptune. There is a chance that Uranus's dark cloud was a vortex.

Uranus's ellipticity value has been a topic of debate for over a century. We understand this characteristic much better today, but further work may be needed. Voyager 2 radio, gravity and imaging data are consistent with an ellipticity (or oblateness) value of 0.0229. This value is a little higher than the ellipticity value measured at the 1 μbar level from stellar occultation data, which was 0.0197.

Another characteristic of Uranus's atmosphere is that it contains two compounds with deuterium, CH_3D and HD. (Deuterium is a special type of hydrogen that contains both a proton and a neutron in its nucleus; it has a symbol of D. The normal hydrogen atom has only a single proton in its nucleus.) The ratio of CH_3D to CH_4 is 3.6×10^{-4} to 1.0. This result is consistent with a D/H ratio of $\sim 7 \times 10^{-5}$ to 1 for Uranus, which is about a factor of five greater than the corresponding value for interstellar hydrogen and about three times the value for Jupiter.

There are two types of molecular hydrogen (H_2) – ortho and para hydrogen. The spins of the two protons in ortho hydrogen are parallel, whereas they are anti-parallel in para hydrogen. The ratio of ortho to para hydrogen can serve as a probe of the dynamics of Uranus's atmosphere. Essentially, the equilibrium ratio of ortho to para hydrogen changes with temperature. For example, at 60 K, the equilibrium ratio is 35% ortho and 65% para, whereas at 140 K, it is closer to 70% ortho and 30% para. If hydrogen at 140 K with 70% ortho hydrogen is injected into the upper troposphere where the temperature is 60 K, much of the ortho hydrogen will cool off and change to para hydrogen so that the new equilibrium ratio is reached. The conversion of ortho to para hydrogen is a slow process. Therefore, if there is lots of hydrogen that is being pushed from the hot interior to the cool upper troposphere through convection, the excess ortho hydrogen will not have enough time to fully convert to para hydrogen and the hydrogen will not be at equilibrium. As it turns out, most of the hydrogen in the upper troposphere and stratosphere is at equilibrium, which is consistent with little or no convection.

Interior of Uranus

We do not have samples of Uranus's interior; therefore, we must use models to predict the nature of its interior. These models are constrained by several characteristics of the planet which include its: mass, radius, temperature, rotation rate, ellipticity, chemical composition and mass distribution within the interior. In addition, interior models of Uranus depend on the equations of state for the compounds making up that planet. The planetary characteristics are summarized below, followed by a discussion of equations of state.

We know the mass of Uranus from both its gravitational tug on Voyager 2 and the movement of its moons.

The radius is also well-known from Voyager 2 data.

We have some idea of the atmospheric temperatures above the two bar level. Astronomers must make assumptions about the temperature profile below this level. In doing this, they make certain assumptions, and use thermodynamics to predict the temperature as the pressure rises inside of Uranus.

Operators of Voyager 2 measured the rotation period of Uranus's interior in 1986.

The ellipticity of Uranus is also known.

We have an idea of the mass distribution inside of Uranus from occultation studies of the rings. Essentially the gravitational field of Uranus is not perfectly symmetrical, and, by combining this asymmetry along with Uranus's rotation rate and ellipticity, one can get an idea of the mass distribution inside of that planet. The mass distribution affects a planet's moment of inertia. Uranus's moment of inertia (I) is $0.24\,MR^2$, where M is the mass and R is the radius of Uranus. This value is much lower than for a sphere of uniform density ($I = 0.40\,MR^2$). This result is consistent with the central layers of that planet being denser than the outer layers.

What is the chemical composition of Uranus? This is a difficult question to answer because we simply do not have much data of that planet's interior. The best that we can do right now is to look at the composition of the Sun and other giant planets and speculate what the bulk composition of Uranus is. Many astronomers believe that Uranus formed from material left over after the Sun formed. Mathematical models along with the physical constants of Uranus are consistent with that planet containing other elements besides hydrogen and helium. Astronomers believe that many of the most abundant elements in our Sun, such as oxygen, nitrogen, and carbon are present in Uranus. That planet does not have as high of a percentage of hydrogen and helium as the Sun because its gravity is too low. Essentially, Uranus was not able to hold on to its light gases very well during its formation phase.

Uranus is assumed to be made up of three components, described as gas, ocean and rock. It is important to realize that the terms "ocean" and "rock" do not necessarily refer to a liquid or solid phase.

The gas component is a mixture of hydrogen, helium, and other gases. The ocean component is probably a fluid layer that contains a mixture of oxygen, nitrogen, carbon, and hydrogen. Most astronomers believe that the ocean layer is made up mostly of water, ammonia, and carbon dioxide. The rocky core component probably contains the compounds silicon dioxide, magnesium oxide, iron sulfide, iron oxide, nickel sulfide, and nickel oxide. The rocky core is believed to be in a liquid phase.

To compute useful models of the interior, the equations of state are required.

The equation of state predicts the density of a material at different pressures. The internal pressure in Uranus can reach a few million bar. The equation of state for water predicts a density of ~ 2 g/cm^3 at a pressure of 100,000 bar, which is twice the density at a pressure of 1.0 bar. The equations of state for hydrogen, helium, and water are fairly well known, but those of the rocky material are not as certain.

Several groups have carried out calculations of Uranus's interior using different assumptions about chemical composition and other model parameters. The main findings of these studies are:

(1) Uranus is made up primarily of other materials besides hydrogen and helium.
(2) Uranus either does not have a rocky core of if it does then the core makes up less than 5% of that planet's mass.

Two possible models of Uranus are shown in Figures 1.13. In the first model, there are three distinct layers of rock, ocean, and gas, while in the second one, the composition is more homogeneous. Uranus is probably closer to the left part or the first model in Figure 1.13, but the boundaries may not be as distinct as within Earth. Recent model calculations suggest that if Uranus has a rocky core then the core has a mass less than that of Earth. The ocean layer contains about 80% of the planet's mass and the gas layer contains about 15–20% of the mass.

Table 1.6 summarizes the amount of solar energy several planets receive along with how much heat their interiors release. The amount of solar energy received is the average amount received per square meter for each planet; both daylight and night hemispheres are included. This evaluation makes the situation the same as that for the amount of internal heat released. Uranus gives off a smaller amount of heat than the other giant planets. One possible explanation for this is that Uranus has stratified layers of material that do not allow much heat to escape. A thermos bottle retains heat in a similar way.

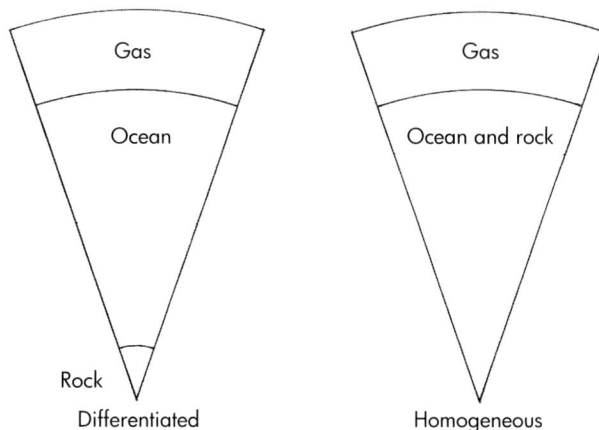

Figure 1.13. Two possible models of the interior of Uranus. The gas, ocean and rock layers are separated from one another on the left. This represents a differentiated interior. The rock and ocean components are mixed with each other on the right producing a homogeneous interior. (Credit: Richard W. Schmude, Jr.)

Table 1.6. A summary of the average amount of heat received from the Sun and the amount of heat given off by the interiors of the five largest planets in our Solar System

Planet	Solar energy received Joules/(m² s)[a]	Internal heat released[b] Joules/(m² s)	Total heat released Joules/s
Earth	342	~0.065	3×10^{13}
Jupiter	12.6	5	3×10^{17}
Saturn	3.76	2	9×10^{16}
Uranus	0.928	~0.04	3×10^{14}
Neptune	0.378	0.4	3×10^{15}

[a] These are average values averaged for both day and night; these values can thus be compared to those corresponding to the release of internal heat in the third column. Internal heat is released over the entire surface including areas on the night side.
[b] Values for Jupiter, Saturn, Uranus, and Neptune are from Cruikshank et al (1995) p. 114. The value for Earth is from Woodhead (2001) p. 49.

Magnetic Environment

Figure 1.14 shows the orientation of Uranus's magnetic field in 1986. The magnetic field has two characteristics, which are: (1) its large tilt from the rotational axis and (2) the fact that it is centered on an area that is far from the center of Uranus. Implications of each of these characteristics are discussed next.

The Uranian magnetic field is inclined to the rotational axis by 59°, which is different than the fields of Earth, Jupiter, and Saturn. The magnetic poles on Uranus are far from the rotational poles. As a result, the auroras develop far from the rotational poles and can face Earth. In addition, a large range of magnetic

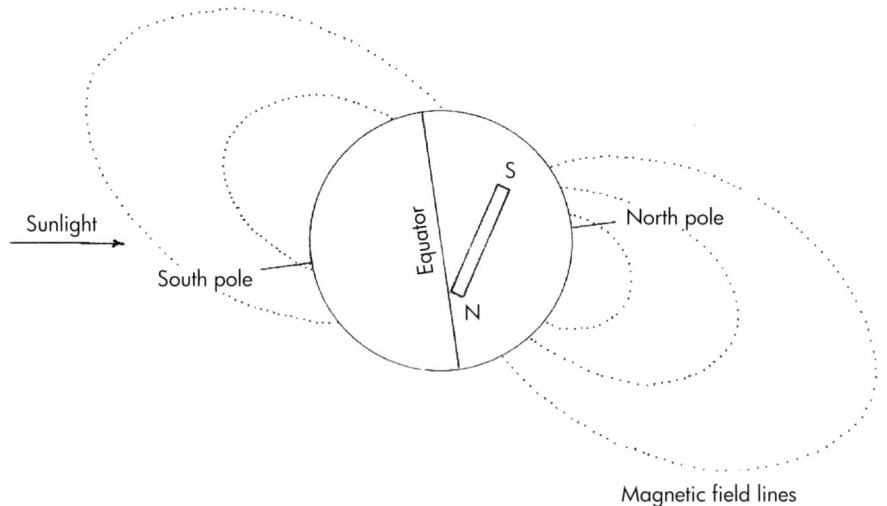

Figure 1.14. A diagram of the positions of the two rotational poles and that planet's magnetic field, which is represented as the bar with N and S at opposite ends. The magnetic field was tilted at an angle of 59° from the rotational axis in 1986. (Credit: Richard W. Schmude, Jr.)

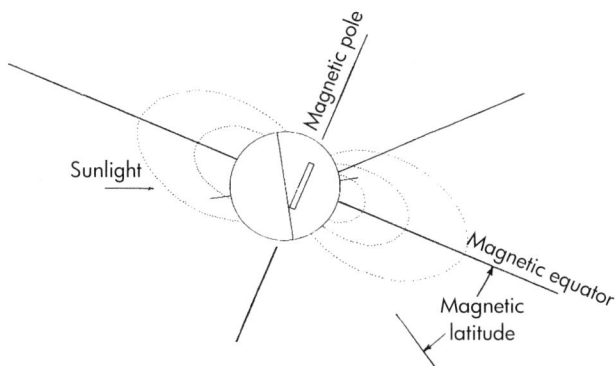

Figure 1.15. The magnetic poles and the magnetic equator. The magnetic equator is at a 90° angle from the line defined by the magnetic poles. The magnetic latitude of an object is the smallest angle between the object and the magnetic equator measured from the center of the magnetic field. (Credit: Richard W. Schmude, Jr.)

latitudes face the Sun each time Uranus rotates. The large tilt of the magnetic field also causes the rings and satellites to sweep through a wide range of magnetic latitudes. Figure 1.15 defines the magnetic latitude for Uranus. The magnetic latitude ranges from 90° at the magnetic pole to 0°, which is half-way between the two magnetic poles.

A second characteristic of Uranus's magnetic field is that it is offset from the planet's center by ~8,000 km (or 0.3 planetary radii). As a result, the magnetic field varies widely across the cloud tops. Its average strength is between 0.2 and 0.3 Gauss (G) at the 1 bar level, but it ranges from 0.1 G up to 1.0 G. For a comparison, the average magnetic field strength at Earth's surface is 0.3 G. Uranus's magnetic field has a more complex force-field geometry than Earth's field. Large quadrupole and octupole components are present in Uranus's field.

For a dipole field, the L value, in units of planetary radii, is:

$$L = r/(cos^2\lambda) \tag{1.4}$$

where r is the distance from the dipole center in units of planetary radii and λ is the magnetic latitude in degrees. Charged particles (electrons and cations) zip along in areas with constant L; hence, one must know the L value in order to study Uranus's magnetic environment. The value of L is the distance from the dipole center at a magnetic latitude of 0°. Figure 1.16 shows magnetic field lines with L = 5, 10 and 15 Uranus radii. For distances exceeding about two planetary radii, one can assume that the magnetic field behaves as a dipole and so one can use equation 1.4, to

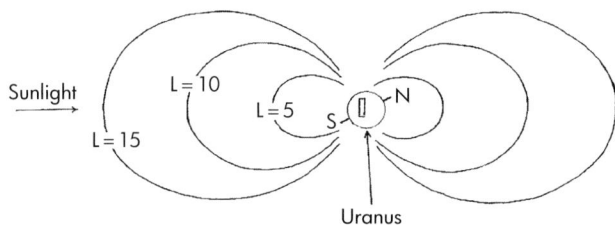

Figure 1.16. Uranus and the L = 5, 10 and 15 magnetic field lines. (Credit: Richard W. Schmude, Jr.)

predict field lines. At close distances, however, this assumption breaks down due to the large quadrupole and octupole contributions.

What causes Uranus to have a magnetic field? Scientists believe that magnetic dynamos generate planetary magnetic fields. In the case of Earth, they believe that the dynamo is caused by the release of heat by radioactive elements which then create convection currents within the core. Earth's rotation may also play some role in the dynamo. Scientists are less sure of Uranus's magnetic field. Like Earth, Uranus probably has some kind of dynamo. There is a good chance that Uranus's dynamo lies in the ocean layer. One recent study suggests that it may be ~8,000 km below the 1.0 bar level.

Uranus's magnetic field extends far into outer space, and deflects solar wind particles as is shown in Figure 1.17. The magnetosphere is the region around the planet where the dominant magnetic field is from the planet and the magnetopause is the outer boundary of the magnetosphere. The area between the bow shock and the magnetopause is a transition area that is neither dominated by the solar wind or Uranus's magnetic field. In 1986 as Voyager 2 passed through this region, both the strength and direction of the magnetic field fluctuated. The magnetopause was ~460,000 km from the center of Uranus (on the Sunward pointing side). The strength of the solar wind, along with the strength and orientation of the planetary magnetic field controls the size and shape of the magnetosphere. Voyager 2 passed Uranus during solar minimum and so the force of the solar wind may have been weaker than normal. The shape of Uranus's magnetosphere may change as the

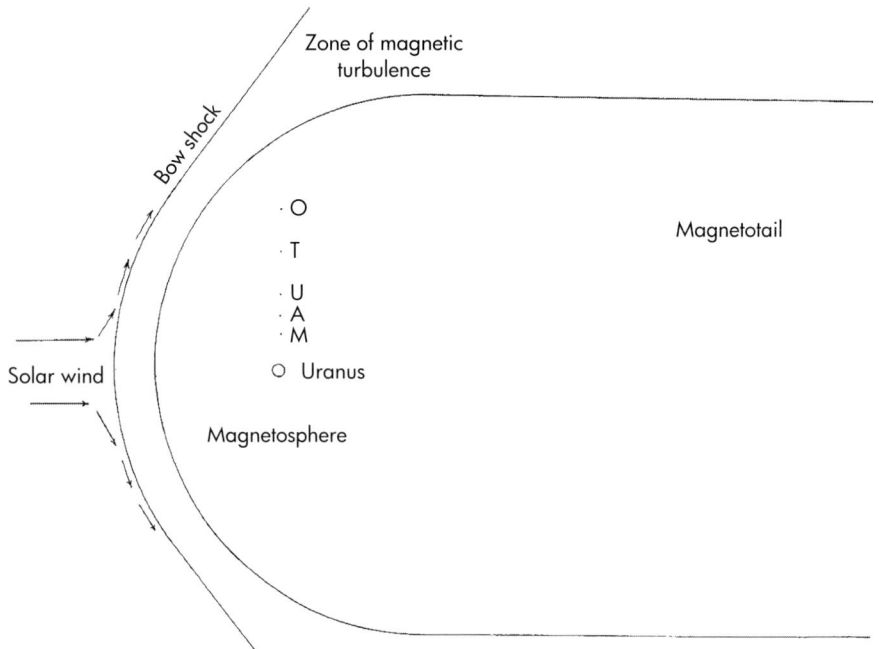

Figure 1.17. Side view of Uranus and its magnetosphere in 1986. Uranus is the small circle. Uranus's magnetic field deflected the solar wind creating the magnetosphere. The points M, A, U, T and O correspond to the distances of Miranda, Ariel, Umbriel, Titania and Oberon from Uranus. As one can see, all of these moons were inside of Uranus's magnetosphere in 1986. (Credit: Richard W. Schmude, Jr.)

planet rotates due to the high inclination of the magnetic field. Finally, there may be seasonal changes in the magnetosphere since the strongest pole (the one closest to Uranus's surface) was on the night side of the planet in 1986. A future probe will be needed to verify these changes.

What lies inside of Uranus's magnetosphere? Both cold and hot plasma lie within its magnetosphere. Cold plasma is more likely to remain trapped in a magnetic field whereas hot plasma is more likely to escape. Charged particles in a cold plasma move at slower speeds than those in a hot plasma. Both types of plasma were in the magnetosphere in 1986. Many of the slower moving plasma particles move in helical paths around magnetic field lines as shown in Figure 1.18. The kinetic energies of plasma particles are often expressed in units of electron volts (eV) kiloelectron volts (keV) and megaelectron volts (MeV).

The distribution of high-energy electrons is shown in Figure 1.19. There are two regions at L = 7.3 and L = 5 where there are drops in electron density. These drops occur when charged particles strike the moon Ariel (minimum L = 7.3) and the moon Miranda (minimum L = 5). Even though Miranda and Ariel cross several L-shell values, they spend much of their time near their minimum L-shell and so absorption is greatest here. The smaller inner moons undoubtedly absorb many charged particles as well.

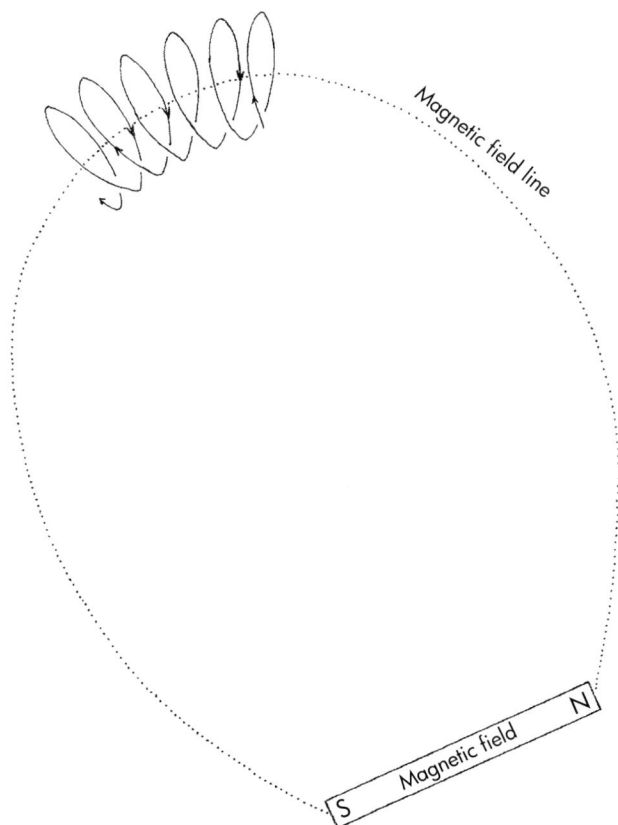

Figure 1.18. A charged particle travels in a helical path around an invisible magnetic field line. The invisible magnetic field line is the dotted curve. (Credit: Richard W. Schmude, Jr.)

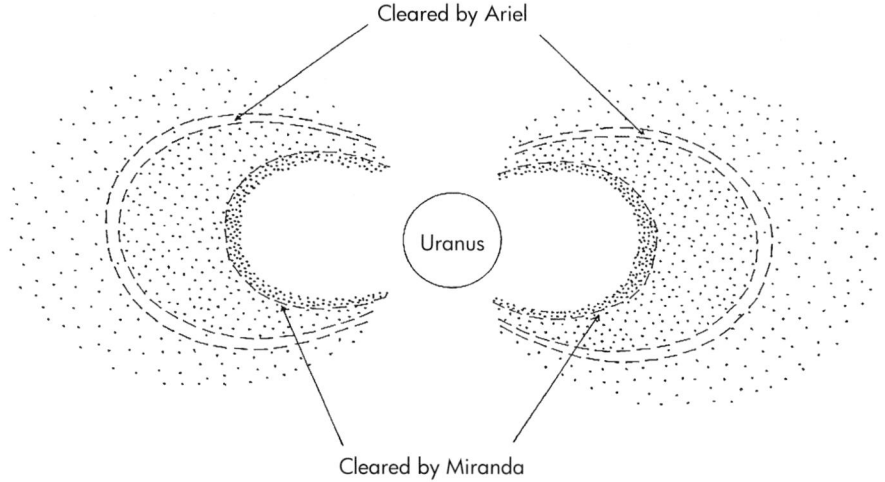

Figure 1.19. Distribution of high-energy electrons at different distances from Uranus. Dots show areas of high electron density. Ariel has almost cleared out one area between the two dashed curves near L = 7.3, and Miranda has reduced the number of electrons near the single dashed curve near L = 5. (Credit: Richard W. Schmude, Jr.)

About 99% of the cations, with energies between 0.6 and 1.0 million electron volts per nucleon (hot plasma) in Uranus's magnetosphere are H^+ ions and $\sim$1% are H_2^+ ions. There are almost no He^+ ions. Since there is almost no He^+, the solar wind (which contains substantial amounts of He^+) is not the principal source of the hot plasma. Large amounts of the ions in the magnetosphere probably come from the ionosphere and corona, where there is little helium.

The magnetotail lies beyond the magnetosphere; it has a cylindrical shape and is at least several million kilometers long. It is just over two million kilometers in diameter. Voyager 2 went through the magnetotail as it passed Uranus. The magnetic field strength dropped more gradually with increasing distance from Uranus in this area than in the magnetosphere. A cross-section view of the magnetotail is shown in Figure 1.20. The top half of the magnetotail has an opposite magnetic polarity from the bottom half. Essentially, the north end of a

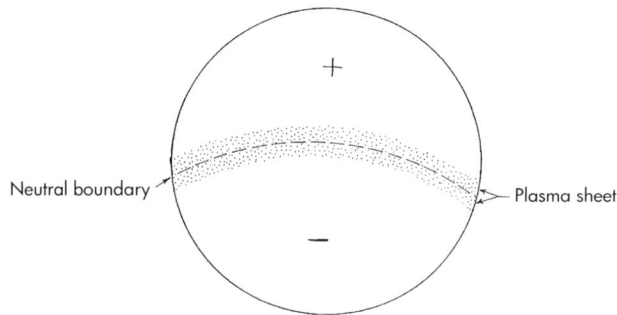

Figure 1.20. This figure shows a cross section view of Uranus's magnetotail. The top part has one magnetic polarity (+) and the bottom part has the opposite magnetic polarity (−). The neutral boundary is the dashed curve and is the area where the polarity changes. The plasma sheet consists of a layer of charged particles near the neutral boundary. (Credit: Richard W. Schmude, Jr.)

compass needle would point in opposite directions in the top and bottom halves. In between these two halves lies the neutral boundary (dashed line in Figure 1.20). This is the area where the magnetic field changes polarity. The area near the neutral boundary is called the plasma sheet (the dotted area in Figure 1.20) and it contains protons and electrons. It rotates at nearly the same rate as Uranus. Voyager 2 passed through Uranus's plasma sheet several times. The plasma sheet has a slight twist. There is some evidence that the sheet is thicker at one end than at the other, and that the thickness increases away from the center of the magnetotail.

In 1986, Uranus emitted large amounts of radio waves having frequencies as low as 10 hertz. Many of these waves came from areas near the magnetic pole in Uranus's northern hemisphere, which was facing away from Earth and the Sun. Some of the radio waves also came from the magnetosphere. Some changed every 17.24 hours, which is the rotation of Uranus's magnetic field. That planet emitted short bursts of radio waves. Lightening bursts below the 1.0 bar level may be the source of these bursts. Voyager 2 recorded over 100 of these events.

If the orientation of Uranus's magnetic field remains nearly constant, the magnetic poles will face Earth several times in the 21st century. These would be good times to study the radio emissions of that planet from Earth and determine if the magnetic field orientation has changed since 1986.

Uranus has an aurora. It is located in the dark (northern) hemisphere, is centered on the magnetic pole near 44°N and has a diameter of ~7,500 km. The aurora gives off some ultraviolet light. What causes the aurora? Low energy electrons from the magnetosphere spiral down magnetic field lines and collide with the hydrogen and other gases in the atmosphere at about the 1 μbar level. These collisions cause the gases to give off ultraviolet radiation, which we see as the aurora.

Rings

Uranus has 13 rings. Their names and characteristics are given in Table 1.7. Figure 1.21 shows several of the rings. Scientists discovered the 6, 5, 4, Alpha (α), Beta (β), Eta (η), Gamma (γ), Delta (δ) and Epsilon (ε) rings from Earth in 1977 using stellar occultation data. Four more rings, the Zeta (ζ), Lambda (λ), R/2003 U 1 and R/2003 U2 rings were discovered later with Voyager 2 and Hubble Space Telescope data.

Astronomers have used several methods to gather data on the rings, namely, Earth-based occultations, Voyager occultations, Earth-based images, Hubble Space Telescope images and Voyager 2 images. The occultation sources are described below and are followed by the three sources of images.

An occultation occurs when an object such as a ring blocks out the light of a star that is much farther away. Earth-based data is very useful because each occultation probes a different longitude of a ring and, when data from several occultations over a period of a decade are combined, one can construct an accurate map of a ring at several longitudes. We are also able to learn the ring diameters and widths from this kind of data. Due to the properties of light and the angular sizes of the occulted stars, however, Earth-based occultation data have resolutions of a few kilometers. This means that astronomers are unable to determine the widths of rings that are less than about four kilometers wide from Earth-based data. Voyager occultation data, on the other hand, have resolutions of under 0.2 km; hence, one can measure widths of rings thinner than four kilometers with this data.

Table 1.7. Characteristics of Uranus's rings

Ring	Distance (km)	Eccentricity	Period (days)	Width (km)	Composition[a]
Zeta (ζ)	37,000 to 39,500	?	0.215 to 0.237	2500	?
6	41,837	0.0010	0.258	1 to 3	LP
5	42,235	0.0019	0.262	2 to 3	LP
4	42,571	0.0011	0.265	2 to 3	LP
Alpha (α)	44,718	0.0008	0.286	4 to 13	LP
Beta (β)	45,661	0.0004	0.295	7 to 12	LP
Eta (η)	47,176	0.0000	0.309	~2[b]	LP
Gamma (γ)	47,627	0.0001	0.314	1 to 4	LP
Delta (δ)	48,300	0.0000	0.320	3 to 7[b]	LP
Lambda (λ)	50,024	~0	0.338	0 to 3	D, LP
Epsilon (ε)	51,149	0.0079	0.349	20 to 95	LP
R/2003 U2	67,300	?	0.527	?	D
R/2003 U1	97,700	?	0.922	~20,000	D

[a] LP = large particles; D = dust
[b] Dimensions are for the bight component only.

For several years now, astronomers have used both Earth-based telescopes and the Hubble Space Telescope to image the rings. One strategy that several people have used is to take images in light that the planet absorbs. This causes Uranus to be dim and as a result, its light is less likely to drown out the feeble light reflected by the rings. One must remember that the rings are too narrow to be resolved in both Earth-based images and those made with the Hubble Space Telescope.

In the case of Voyager images, astronomers were not as concerned with light from Uranus; however, they were concerned with the rapid movement of the spacecraft. In many cases, Voyager imaged only a small portion of the rings. With the exception of the Epsilon ring, the rings were generally too narrow to be resolved in Voyager images.

We know that the six rings (Alpha, Beta, Eta, Gamma, Delta, and Epsilon) reflect about the same amount of violet and green light. We also know that in visible light, the rings are made up of low albedo particles. In addition, some have imaged the rings in near-infrared light. These images show that the nine rings discovered in 1977 change in brightness at different longitudes. For most of these rings, we are not sure whether their widths change with longitude or if they have brighter parts at some longitudes.

Earth passed through the ring plane in May and August of 2007. During these times, astronomers imaged the rings with several ground-based instruments. Astronomers report that the faint outer ring, R/2003 U1, was more distinct in the ring plane crossing images than in images made in previous years.

One obvious question is: where did the rings come from? One clue is the fact that all but one or two lie within the Roche zone. This zone extends out to about 68,000 km from Uranus's center for objects with a density of 1.0 g/cm^3. Any object that is held together by just its own gravity will be torn apart by tidal forces if it enters the Roche zone. A small object held together by chemical bonds (like a space

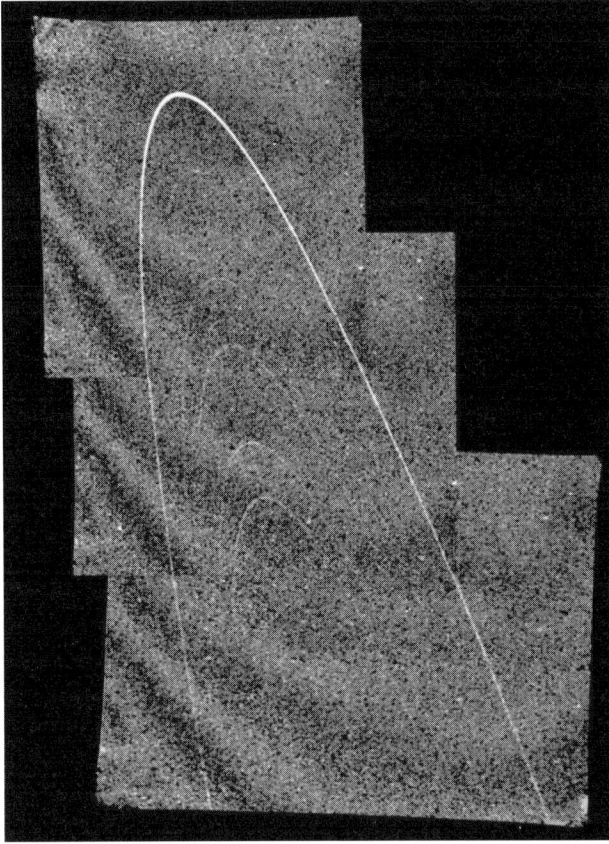

Figure 1.21. A Voyager 2 image of some of the rings taken in backscattered light. The thin bright arcs are the rings. (Credit: Courtesy NASA/JPL-Caltech.)

probe or a car-sized boulder) will, however, not be torn apart. Astronomers believe that it is not possible for a large object to form from smaller ones inside of the Roche zone. The rings, therefore, may have come from either an object that broke apart inside of the Roche zone or from material that never assimilated into a moon. Recent images are consistent with the small moon Mab being a source of particles for the newly discovered ring R/2003 U 1. One or more small undiscovered moons may be a source of particles for the newly discovered ring, R/2003 U 2.

One possible scenario for the development of many of the rings is that a small satellite within the Roche zone broke apart. This break-up may have been hastened by the impact of a large object. The satellite fragments then spread out around the planet. As the particles moved around it, they collided with one another creating more fragments. These collisions caused the rings to become both flatter and wider. Particles on the inner edge of the ring drifted closer to the planet, and those on the outer edge drifted farther from the planet. As this was happening, the small particles lost enough momentum from gas drag forces that they spiraled into Uranus. Finally, gravitational forces from one or more small satellites prevented the rings from continuing to spread farther. Erosion of ring particles continues to the present, and is a source of dust in and near the rings.

Soon after the rings were discovered, astronomers began asking "what forces keep the rings together?" People realized that ring particles are exposed to several forces including gravity (from the planet, moons, rings and Sun), gas drag from Uranus's corona, collisions between ring particles, meteoroid collisions, interactions with sunlight called the Poynting-Robertson effect and possibly electrostatic forces from Uranus's magnetosphere. (The Poynting-Robertson effect occurs when ring particles absorb sunlight and then preferentially re-emit the light in a forward direction. The net effect of this is that particles move in a certain direction as a result of Newton's third law.) As mentioned earlier, rings tend to spread apart unless some force prevents it. The most likely force keeping the rings in their narrow orbits is gravity from small moons. Essentially, when a satellite is at certain positions, it can exert gravitational tugs on a ring. The two small moons Cordelia and Ophelia probably interact with a few of the rings.

Until 1986, astronomers had little information about the nature of the ring particles and the thickness of the rings. This is due to the fact that the solar phase angle of Uranus is always $3°$ or less; hence, we are only able to see backscattered light from the rings. Backscattered light occurs when light hits a target and is reflected backwards, whereas forward-scattered light continues in the same direction. This is illustrated in Figure 1.22.

How can astronomers determine the size of the particles making up Uranus's rings? In order to answer this question, I will need to discuss how particle size and the wavelength of light affect forward-scattered and backscattered light.

Objects that are much larger than a wavelength of electromagnetic radiation will appear bright in backscattered light but dim in forward-scattered light. (Examples of electromagnetic radiation include: radio, infrared, visible and ultraviolet waves; this type of radiation is discussed further in Chapter 6.) The situation is different for objects that have sizes comparable to the wavelength of electromagnetic radiation being studied. In this case, these objects will be bright in forward-scattered light but dim in backscattered light. If, for example, one is studying how particles affect visible light (wavelength of 0.55 µm) dust would be brighter in forward-scattered light than in backscattered light. This is because dust particles have sizes comparable to the wavelength of visible light. One-meter boulders, however, would be brighter in backscattered light than in forward-scattered light. Therefore, astronomers can determine the size of the particles making up Uranus's rings by measuring the brightness of the rings in backscattered and

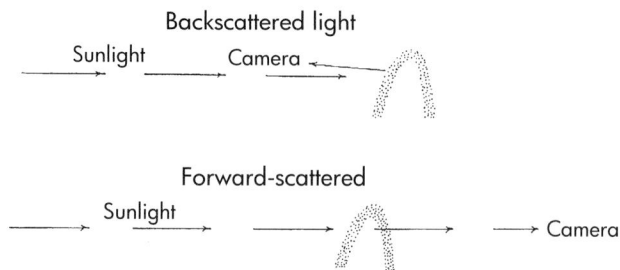

Figure 1.22. Backscattered light is shown in the top part of the figure. Essentially, light is reflected back to its source. Forward-scattered light is light that moves forward towards the camera. Dust is bright in forward-scattered visible light whereas large objects are bright in backscattered light in visible light. (Credit: Richard W. Schmude, Jr.)

forward-scattered light. Unlike Earth-based observers, a space probe can study Uranus's rings in both backscattered and forward-scattered light.

The nine rings discovered from Earth were brighter in backscattered visible light than in forward-scattered visible light, which shows that they are composed mostly of large particles. During occultations, these rings blocked out about equal amounts of radio waves, infrared light and ultraviolet light; this is also consistent with their containing mostly large particles. One reason for the lack of dust may be that it spirals towards Uranus as a result of drag forces from the corona. The corona does not affect large ring particles as much as smaller ones.

In the next sections, each of the rings, starting with the Epsilon ring, is discussed.

Epsilon Ring

During the past few years, astronomers have learned several of the characteristics of the Epsilon ring which are summarized below. This ring reflects more visible light than all of the other rings combined. Its mass density is between 25 and 80 grams per square centimeter. This means that 1.0 square centimeter of the ring has a mass of 25–80 grams. From an assumed mass density of 50 grams/cm^2, I compute a total mass of 9×10^{15} kg for this ring, which is equal to the mass of a 23 km spherical moon with a density of 1.5 g/cm^3. The Epsilon ring also has a width that changes with longitude. At one end, it is only $\sim$20 km across while at the other, it is $\sim$95 km across. See Figure 1.23. The particle density also changes with

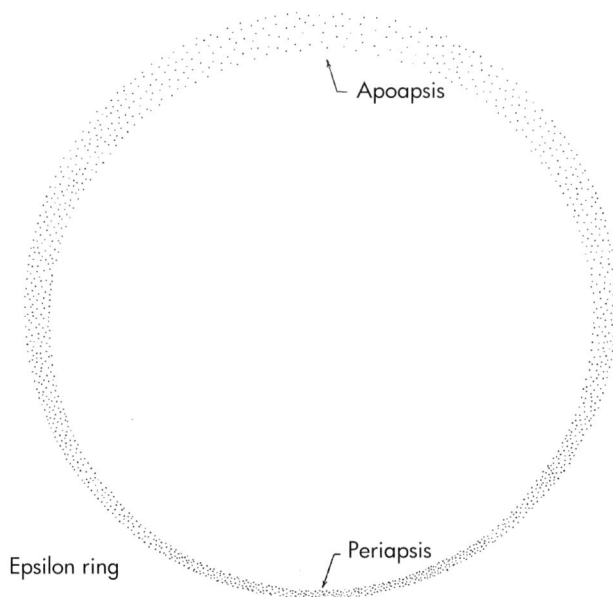

Figure 1.23. Periapsis is the point where the ring is closest to Uranus and apoapsis is the point farthest from it. The epsilon ring is narrow near periapsis and wide near apoapsis. The particles appear closer together at periapsis either because the ring is deeper here or because the particles are closer to each other at this location. (Credit: Richard W. Schmude, Jr.)

longitude. At the narrow end, the particles are either closer together or the ring is deeper than at the wide end. Astronomers also know that the Epsilon ring is inclined to Uranus's equatorial plane. The two points where the ring intersects Uranus's orbital plane are called nodes and these two points define the line of nodes. Because of the inclination of the rings and the fact that Uranus's gravity is not quite symmetrical, the ring undergoes precession. This is illustrated in Figure 1.24. Essentially, as the ring particles move around Uranus once every ~8.4 hours, the elliptical shape of the ring also gradually moves around the planet. It takes the line of the nodes 264 days to make one full trip around Uranus. The rate of precession gives us information about Uranus's gravitational field.

Particles in the Epsilon ring move in an eccentric orbit, which means that this ring is not exactly centered on Uranus. The portion of the ring that is closest to Uranus is called periapsis (or periapse) and the portion that is farthest from Uranus is called apoapsis (or apoapse). See Figure 1.23. The narrow part is close to periapsis and the wide part is close to apoapsis.

The particles making up the Epsilon ring are dark. The particles reflect no more than 6% of the visible light falling on them. This is not the case for Saturn. Most of the ring particles orbiting that planet reflect about 60% of the visible light striking them. The low albedo of Uranus's ring particles is one reason why it is so difficult to see or image the rings. In fact, when the Epsilon ring is almost wide open (near Uranus's solstice) it is less than one one-thousandth as bright as Uranus in visible light. Why are the ring particles so dark? One reason may be that at one time they contained lots of frozen methane and that magnetospheric particles destroyed the methane leaving behind carbon, which is dark. Alternatively the ring particles may be made up of dark carbon-based material that has remained unaltered since the early stages of the Solar System.

Voyager 2 occultation data are consistent with a maximum thickness of 150 meters for the Epsilon ring. Several studies show that the mean particle size is around 0.7 meters, and that each particle is on average about 5 diameters from

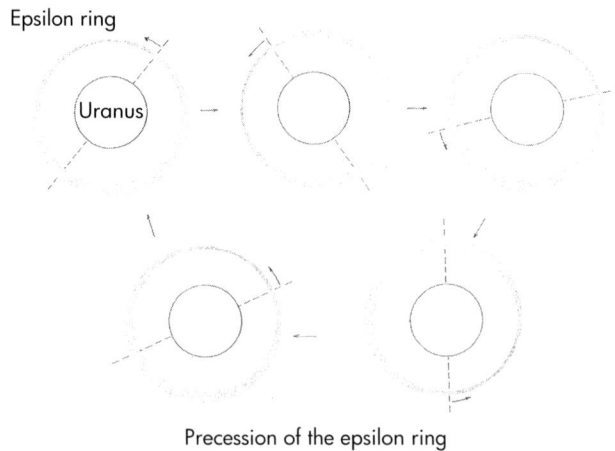

Epsilon ring

Uranus

Precession of the epsilon ring

Figure 1.24. Ring precession. The dashed line represents the line of the nodes and is the intersection between the equatorial plane of Uranus and the plane containing the epsilon ring. The line of the nodes makes one complete trip around Uranus in about 264 days. This movement is called precession. (Credit: Richard W. Schmude, Jr.)

the next one. These studies also show that this ring is not a monolayer of particles but, instead, is several particles thick. Accordingly, a thickness of 50–100 meters is adopted. Keep in mind, though, that the thickness may not be uniform across the entire ring.

Lambda Ring

The Lambda ring is made up mostly of particles with diameters of less than 1 μm. This ring has a lower density of material than the Epsilon ring, and as a result, it does not block out much light when it moves in front of a star. This is why it was not discovered from Earth-based occultation studies. Voyager 2 occultation data show that this ring does not have sharp edges. It also had brighter areas at some longitudes in a forward-scattered light image made by Voyager 2. The brighter areas may be dust coming off of large particles or just clumps of material. An occultation of a star by the Lambda ring on July 11, 1992, is consistent with parts of this ring being much denser than other parts. In fact, the ring may be discontinuous at some longitudes. There is some evidence that this ring, like the Epsilon ring, varies in width at different longitudes, but more data are needed for confirmation.

Delta Ring

The Delta ring has two components; a narrow component and a wide component. The narrow component is brighter than the wider one and it has a width that changes with longitude like the Epsilon ring. It also has a sharp outer edge, which is probably caused by the moon Cordelia. Essentially, for every 22 trips around Uranus that Cordelia makes, the outer edge makes 23 trips. Many astronomers believe that this is what prevents the Delta ring from expanding outwards. The diffuse component has a width of 10 km which is greater than the width of the narrow component. Astronomers believe that both components of the Delta ring are dominated by large particles. One group has reviewed much of the Earth-based occultation data and they estimate a surface density of 7 grams/cm^2 for the narrow component. A five-kilometer object with a density of 1.5 g/cm^3 would be sufficient to create the narrow component.

Voyager occultation data show evidence of a pattern of bright ringlets inside of the Delta ring. One group believes that they are density waves caused by an unseen moon orbiting inside of this ring.

Gamma Ring

The Gamma ring is narrow and, hence, Earth-based occultation studies were unable to yield information about width changes at different longitudes. Voyager occultation data are consistent with a changing width at different longitudes; furthermore, Keck II images show evidence of a changing brightness at different longitudes, which is consistent with a varying width. This ring has sharp inner and outer edges. The moon Ophelia may be responsible for the sharp inner edge because it revolves around Uranus almost five times for every six times that the inner edge particles revolve.

Eta Ring

Like the Delta ring, the Eta ring has two components: a wide component and a narrow one. The narrow component covers the inner part of this ring and is brighter than the wide or outer component. Its width is near two kilometers. During one Voyager occultation measurement, the narrow component did not block out any star light, which is consistent with it not being complete; however, it blocked out starlight many times in Earth-based studies. This ring is the only one that is circular with almost no inclination. The outer, diffuse component is about 55 km wide, but it blocks out almost no starlight during an occultation. The outer component has a lower density of particles than the narrow component.

Beta Ring

The Beta ring width ranges from about 7 to 12 km and changes with longitude. Voyager 2 occultation data show that this ring lacks sharp inner and outer edges. Two scientists estimate a surface mass density of 1.5 gram/cm^2 and a ring mass of 3.8×10^{13} kg.

Alpha Ring

The Alpha ring is similar in size and width to the Beta ring. It has a width that changes with longitude. The minimum width is 4 kilometers and the maximum width is 13 kilometers. Like the Beta ring, it also lacks sharp borders. The Alpha ring has an estimated mass of 4×10^{13} kg.

Rings 4, 5, and 6

These three rings are too narrow to resolve in Earth-based studies. Voyager occultation data along with Earth-based images are consistent with them having average widths of about two kilometers but that the widths probably change with longitude. Each of these rings has estimated masses of 10^{13} kg.

Zeta Ring

Scientists imaged the Zeta ring in images made with the Keck II telescope in near-infrared light. This ring is wide but it has a low particle density. As a result of this, it did not block out any measurable light in Voyager and Earth-based occultations.

Part of this ring was probably imaged by Voyager 2 in visible light at a phase angle of 90°. The ring in the Voyager images (called 1986U2R in 1986) was narrower than the Zeta ring. This difference may be due to a number of factors including the different wavelengths of light used or the difference in the solar phase angles used in Voyager and Earth-based images.

If the Zeta ring moves in front of a star, then over 99.9% of the light would pass through. This is consistent with the low particle density in this ring.

New Rings

In 2005, astronomers discovered two new rings, R/2003 U1 and R/2003 U2, around Uranus. Subsequently they searched for these rings in previous images and found them in a 2003 Hubble image and in 1986 Voyager 2 images.

The new ring, R/2003 U1, has a peak density at the orbit of the moon Mab. It is made up mostly of small dust particles. It is unique among Uranus's rings because of its large distance from its primary – Uranus. This ring lies beyond Uranus's Roche zone. Dust in this ring will experience a lower drag force because the corona gets thinner with increasing distance from Uranus.

A second new ring, R/2003 U 2, lies outside of the Epsilon ring. It is also made up of dust. A small unseen moon (or moons) is believed to be the source of dust for this ring. We are not sure if this ring lies in Uranus's Roche zone. As of 2006, we know very little about the shapes and widths of both R/2003 U1 and R/2003 U2.

Beyond the Ring System

Voyager 2 crossed Uranus's ring plane at a distance of ~115,000 km from the planet's center. As it crossed the plane, it struck dust particles at rates exceeding 50 particles per second. This data, along with the velocity of the probe, allowed scientists to estimate that there is only about one dust particle per 1,000 cubic meters of space at the distance where Voyager 2 passed. A large house has a volume of about 1,000 cubic meters. A dust sheet with this low density would remain invisible from Voyager 2 and Earth-based images. A similar dust sheet lies outside

Figure 1.25. Voyager 2 image of the rings in forward-scattered light. Many of the bright bands are not visible in backscattered light, which shows that the bands contain lots of dust with few large objects. (Credit: Courtesy NASA/JPL-Caltech.)

of Saturn's rings, but is about 1,000 times denser than the one around Uranus. The narrow rings are enclosed in a thin envelope of dust. The dust is bright in forward-scattered light. See Figure 1.25. This envelope has a low dust concentration. As Earth crossed the ring plane in 2007, astronomers imaged this dust. They report that it has changed since 1986.

Satellites

Uranus has 27 known moons. Table 1.8 lists their names, orbital and physical characteristics; and Table 1.9 lists their photometric constants. The moons are broken down into three categories: regular satellites, collision fragments and captured objects. Each category is discussed, and individual moons in each category are also described.

What are the characteristics of a regular satellite? First of all, this satellite has a nearly circular orbit with a low inclination and has enough mass to force itself into a nearly spherical shape. An object with a diameter of 400–800 km is usually large enough to force itself into a spherical shape. A regular satellite is believed to have formed at about the same time as the planet that it orbits. The five regular satellites of Uranus are Titania, Oberon, Umbriel, Ariel and Miranda.

Like Uranus, the five regular moons experience an 84-year cycle of seasons. During the late 1980s, the southern hemispheres of these moons faced the Sun, but by 2030, their northern hemispheres will face the Sun. There is a chance that some of the moons may grow brighter or dimmer as a result of the northern and southern hemispheres reflecting different amounts of light.

The only close-up images of Uranus's moons that we have were made by Voyager 2. This probe imaged the five largest ones and almost a dozen smaller ones. Due to the fact that Uranus is tipped on its side, plus the fact that these moons lie close to Uranus's equatorial plane, more than half of the surfaces of the five large moons remain unmapped. Furthermore, most of the mapped areas are in the southern hemispheres of these moons. Astronomers have constructed maps for about one-fourth of the surfaces of Titania, Oberon, Umbriel and Ariel. In addition, most of the mapped areas are on the Uranus-facing hemisphere. Despite these limitations, astronomers have some idea of the geological processes that have occurred on these moons.

We are able to obtain topographic information of the five moons by measuring the lengths of shadows, studying stereo images, measuring limb profiles and using a technique called photoclinometry. Photoclinometry is a procedure whereby one obtains topographic information from the different kinds of shadows on features. For example, if sunlight strikes all sides of a hill, the side facing away from the Sun will be dimmer than the side facing the Sun. This is because the shadow casted by an individual dirt particles will be smaller on the side of the hill facing the Sun than if this same particle was on the side facing away from the Sun. Although an individual shadow is too small to be imaged at a distance of several thousand kilometers, the effect of millions of these shadows will cause the side facing away from the Sun to appear dimmer.

Voyager 2 images are consistent with the regular moons having synchronous rotations. This means that they rotate at the same rate that they circle the planet. Essentially, the same side of each moon faces Uranus all of the time. See Figure 1.26.

Table 1.8. Names, orbital and physical characteristics of the moons of Uranus

Name	Distance[a] (km)	Orbital Period (days)	I (°)[b]	Radius (km)	Mass[e] (10^{18} kg)	Density (g/cm^3)
Cordelia	49,600	0.334	0.1	25×18[c]	0.05	1.5?
Ophelia	54,200	0.381	0.1	27×19[c]	0.06	1.5?
Bianca	59,300	0.435	0.2	32×23[c]	0.1	1.5?
Cressida	61,800	0.464	0.0	46×37[c]	0.4	1.5?
Desdemona	62,700	0.474	0.1	45×27[c]	0.2	1.5?
Juliet	64,400	0.493	0.1	75×37[c]	0.6	1.5?
Portia	66,100	0.513	0.1	78×63[c]	2	1.5?
Rosalind	69,900	0.558	0.3	36[c]	0.3	1.5?
Cupid	74,400	0.613	0.0	6[d]	0.001	1.5?
Belinda	75,300	0.624	0.0	64×32[c]	0.4	1.5?
Perdita	76,400	0.638	0.1	13[d]	0.01	1.5?
Puck	86,000	0.762	0.3	81[d]	3	1.5?
Mab	97,700	0.922	0.1	8[d]	0.003	1.5?
Miranda	129,900	1.413	4.3	236[f]	66[f]	1.2
Ariel	190,900	2.520	0.0	579[f]	1270[f]	1.6
Umbriel	266,000	4.144	0.1	585[f]	1290[f]	1.5
Titania	436,300	8.706	0.1	789[f]	3480[f]	1.7
Oberon	583,500	13.463	0.1	761[f]	3020[f]	1.6
Francisco	4,276,000	267.0	145	5[d]	0.0008	1.5?
Caliban	7,231,000	579.7	141	25[d]	0.1	1.5?
Stephano	8,004,000	683.9	144	8[d]	0.003	1.5?
Trinculo	8,504,000	748.9	167	7[d]	0.002	1.5?
Sycorax	12,179,000	1288	159	52[d]	0.9	1.5?
Margaret	14,345,000	1641	57	5[d]	0.0008	1.5?
Prospero	16,256,000	1979	152	13[d]	0.01	1.5?
Setebos	17,418,000	2195	158	13[d]	0.01	1.5?
Ferdinand	20,901,000	2886	170	7[d]	0.002	1.5?

[a] Distance = the average distance or semi-major axis of the orbit.

[b] I = inclination.

[c] Radius values are from Karkoschka (2001b).

[d] Radius values are computed from the V(1,0) and albedo values in Table 1.9 and the equation listed in Grav et al (2004).

[e] Mass values for the satellites with two radius values, like Cordelia, were computed from the density in the far right column and a tri-axial ellipsoid geometry with the third dimension being equal to the smaller number in the radius column. Mass values for the satellites with just one radius value, like Rosalind, were computed from the density value in the far right column and a spherical geometry.

[f] Radius values are from Bergstralh et al (1991) p. 516. For Ariel and Miranda, the average radius is listed. The satellite masses are the average of the three values reported in the same source on p. 521. The writer computed the densities from the selected masses and radius values.

Our moon moves in the same way. The longitudes of the five largest moons are defined as: 0° is the longitude that faces Uranus, 90° is the center of the leading hemisphere, 180° is the longitude opposite from Uranus, and 270° is at the center of the trailing hemisphere.

Table 1.9. Photometric constants of Uranus's moons V_o is the magnitude at average opposition distance and $V(1,0)$ is the magnitude when the object is 1.0 au from the Earth and Sun

Name	V_o	$V(1,0)$	Albedo[i] 0.54 μm	Albedo[i] 0.72 μm	Albedo[i] 0.91 μm
Cordelia[a]	23.6	10.9	~0.1	~0.1	~0.1
Ophelia[a]	23.2	10.5	~0.1	~0.1	~0.1
Bianca[a]	22.5	9.8	~0.1	~0.1	~0.1
Cressida[a]	21.6	8.9	~0.1	~0.1	0.13[d]
Desdemona[a]	22.0	9.3	~0.1	~0.1	0.09[d]
Juliet[a]	21.1	8.4	~0.1	~0.1	0.10[d]
Portia[a]	20.4	7.7	0.11[c]	0.12[c]	0.12[d]
Rosalind[a]	21.8	9.1	~0.1	~0.1	0.12[d]
Cupid[g]	25.3	12.6	~0.1	~0.1	~0.1
Belinda[a]	21.5	8.8	~0.1	~0.1	0.08[d]
Perdita[a]	23.7	11.0	~0.1	~0.1	~0.1
Puck[a]	19.7	7.0	0.10[c]	0.10[c]	0.14[d]
Mab[g]	24.8	12.1	~0.1	~0.1	~0.1
Miranda[a,b]	15.79	3.08	0.38	0.31	0.43
Ariel[a,b]	13.70	0.99	0.48	0.37	0.50
Umbriel[a,b]	14.47	1.76	0.20	0.19	0.25
Titania[a,b]	13.49	0.78	0.34	0.23	0.36
Oberon[a,b]	13.70	0.99	0.29	–	0.32
Francisco[i]	25.6	12.9	0.15?	0.15?	–
Caliban[e]	21.9	9.2	0.15?	0.15?	–
Stephano[e]	24.4	11.7	0.15?	0.15?	–
Trinculo[e]	24.6	11.9	0.15?	0.15?	–
Sycorax[f]	20.3	7.6	0.15?	0.15?	–
Margaret[i]	25.3	12.6	0.15?	0.15?	–
Prospero[e]	23.3	10.6	0.15?	0.15?	–
Setebos[e]	23.3	10.6	0.15?	0.15?	–
Ferdinand[i]	24.8	12.1	0.15?	0.15?	–

[a] The $V(1,0)$ value is the avgerage of V_{max} and V_{min} from Karkoschka (2001a).
[b] Albedos at 0.54 μm are from Bergstralh et al (1991), pp. 536 adjusted for a "Titania-like opposition effect".
[c] The values at a solar phase angle of 1° as reported by Karkoschka were multiplied by 1.1 to account for the expected brightness surge at 0°.
[d] The values at a solar phase angle of 1° as reported by Karkoschka were multiplied by 1.3 to account for the expected brightness surge at 0°.
[e] The $V(1,0)$ values are from Grav et al (2004).
[f] The $V(1,0)$ valule is the average value from Grav et al (2004); Maris et al (2001) and Romon et al (2001).
[g] The $V(1,0)$ value was computed by first adding 0.5 to the R magnitudes reported in IAU circular 8209 and then correcting for the phase angle in the same way as was done by Grav et al (2004).
[h] The $V(1,0)$ value was computed by first adding 0.5 to the R magnitudes reported in IAU circular 8194 and then correcting for the phase angle in the same way as was done by Grav et al (2004).
[i] The $V(1,0)$ value was computed by first adding 0.5 magnitudes to the Aug. 13 R magnitudes in Kavelaars et al (2004) and then correcting for the phase angle in the same way as was done by Grav et al (2004).
[i] All albedos are geometric albedos.

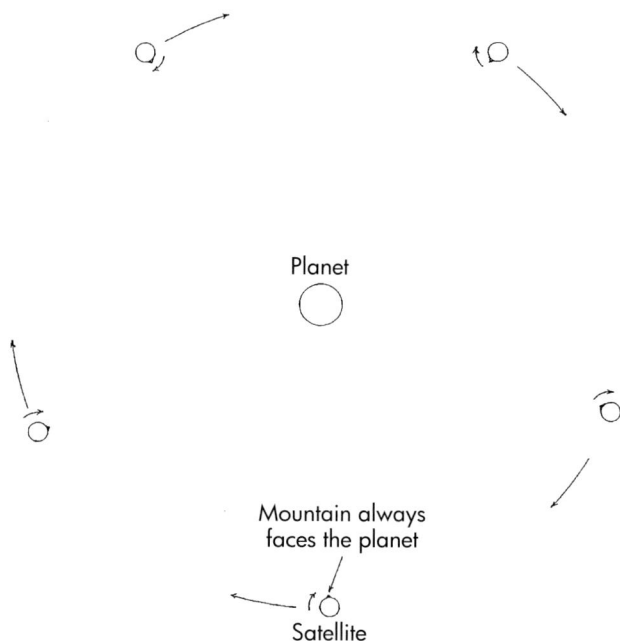

Figure 1.26. The same side of a moon always faces the planet it orbits during synchronous rotation. In this case, a moon's period of revolution around the planet equals its rotation period. The mountain would be at a longitude of 0° since it faces Uranus. (Credit Richard W. Schmude, Jr.)

Voyager 2 gave us information about the sizes and masses of the five regular moons. From this, astronomers were able to compute their densities. The densities range from 1.2 g/cm^3 for Miranda to 1.7 g/cm^3 for Titania. The maximum internal pressure inside the largest moon, Titania, is several thousand bar. This is not enough to create large changes in density. In other words, a material on the surface of Titania will have nearly the same density as if it was at that moon's center. The density of quartz, a common silicate material, is 2.65 g/cm^3, whereas the density of the most primitive carbonaceous chondrite meteorites is $\sim$2.2 g/cm^3, and the density of pure water ice is 0.934 g/cm^3 at 93 K (the approximate temperature of Uranus's moons). Frozen methane, carbon dioxide and nitrogen all have densities below that of water ice. From the densities of the materials just listed, we can conclude that the regular moons contain additional materials besides ices.

The interiors of the large moons are poorly understood. They may have a purely homogeneous mixture of ice, carbonaceous carbonate material and possibly rock, or a stratified (or differentiated) interior composed of different layers of material. See Figure 1.27. A fair amount of internal heat is needed to force the moons to have differentiated interiors. Three sources of internal heat are tidal heating, accretion heating and the decay of radioactive elements.

Tidal heating can occur when a gravitational force from a nearby body causes periodic changes in a moon's shape, which then leads to friction and internal heat. Jupiter's moon Io experiences tidal heating and this is probably why it has active volcanoes.

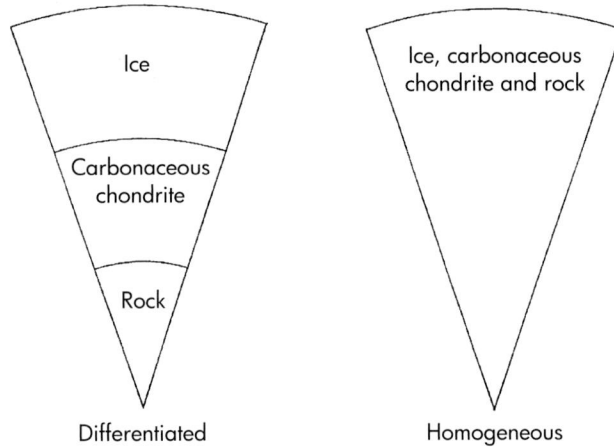

Figure 1.27. Uranus's regular moons may have distinct layers of materials like ices, carbonaceous chondrite material and rock as is shown in the left diagram, or may have a homogeneous composition like what is shown at the right. (Credit: Richard W. Schmude, Jr.)

Many astronomers believe that Uranus's regular satellites formed as a result of accretion. During accretion, dust, rocks and boulders collide with a large object causing it to grow. During these collisions, part of the kinetic energy is transferred to heat. As a result of this, the growing satellite heats up. This heat is called accretional heat. Accretional heating may have helped play a role in the resurfacing of the large moons of Uranus.

Heating can also occur when radioactive elements decay. Essentially, radioactive atoms eject sub-atomic particles at very high speeds which then cause nearby atoms to move. The greater movement of nearby atoms leads to higher temperatures. Rocky material is more likely to have radioactive material than less-dense icy material.

Several of the regular satellites show evidence of volcanic activity. This may seem unreasonable given their small sizes. It is possible for water ice to melt at temperatures far below 273 K. A clathrate can form if one or more substances dissolve in ice. Essentially a clathrate is a solid phase material that contains one or more impurities. The impurities change the bonding in the solid, which leads to changes in the melting point. Some methane-water ice clathrates have melting points below 180 K. There are also ammonia-water ice mixtures that have melting points far below 273 K. Obviously less heat is needed to melt a methane-ice clathrate and certain ammonia-water ice mixtures than pure water ice. Astronomers have not detected any clathrates on Uranus's moons; however, one group may have detected some ammonia hydrate on Miranda.

A brief overview of each of the regular satellites is given.

Titania

Titania is the largest and most massive of Uranus's moons. It is a little less than half the size of our moon and is the eighth largest one in the Solar System. Voyager 2 took close-up images of Titania and two of these images are shown in Figures 1.28 and 1.29. Each pixel in Figure 1.29 is 3.4 km across. It is difficult to

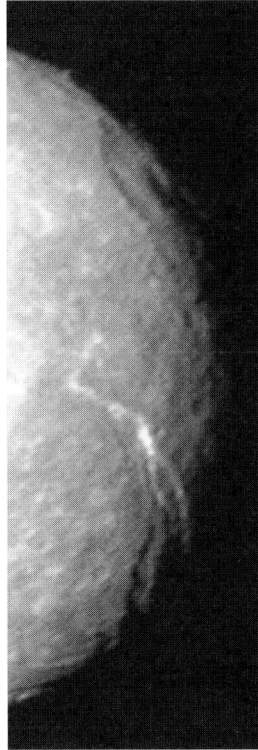

Figure 1.28. Voyager 2 image of the terminator of Titania made on Jan. 24, 1986. There is a large canyon in the lower right and a giant basin at the upper right corner of the image. (Credit: Courtesy NASA/JPL-Caltech.)

Figure 1.29. Voyager 2 image of Titania made on Jan. 24, 1986 at a distance of 500,000 km. One can identify features as small as seven kilometers in this image. (Credit: Courtesy NASA/JPL-Caltech.)

identify craters that are less than about two pixels across; therefore, craters less than seven kilometers are difficult to identify in this figure.

Titania has simple and complex craters along with basins and catenas (crater chains). A simple crater is usually smaller than about 10–20 km and it has both a raised rim and a bowl-shaped appearance. A complex crater is larger than a simple one and has a flat floor and either a central peak or a central pit. The largest crater is a basin. It has a flat floor and can have several nearly concentric rings around it.

Titania also has a few catenas. Similar features are on Ganymede and are thought to be the result of an object that broke apart into several pieces before impact and each fragment created a separate crater. The catenas on Titania may have formed in a similar way or may be the result of faulting or secondary impacts. Secondary impacts occur when debris hurled from one crater hits the surface creating additional craters.

Astronomers counted the number of craters in several areas on Titania and computed the crater density, which is the number of craters per million square kilometers. Crater densities give us information about ages. The older a surface is the higher will be its crater density. The plains near the crater Ursula have a low crater density and an unnamed basin near Titania's equator has a high crater density. Several parts of the cratered terrain have an intermediate density. These results are consistent with relative ages of the unnamed basin being the oldest, followed by the cratered terrain and the plains near Ursula being the youngest. As it turns out, the crater density in Titania's cratered terrain is lower than on Umbriel and Oberon, which is evidence that most of Titania's surface is younger than these two moons.

Figure 1.30 shows a map of part of Titania. The two most distinct craters are Gertrude (diameter = ~300 km) and Ursula (diameter = ~140 km). The crater Calphurnia has a strange central peak which is much longer in one dimension than in the other. Much of the surface in Figure 1.30 is called cratered terrain.

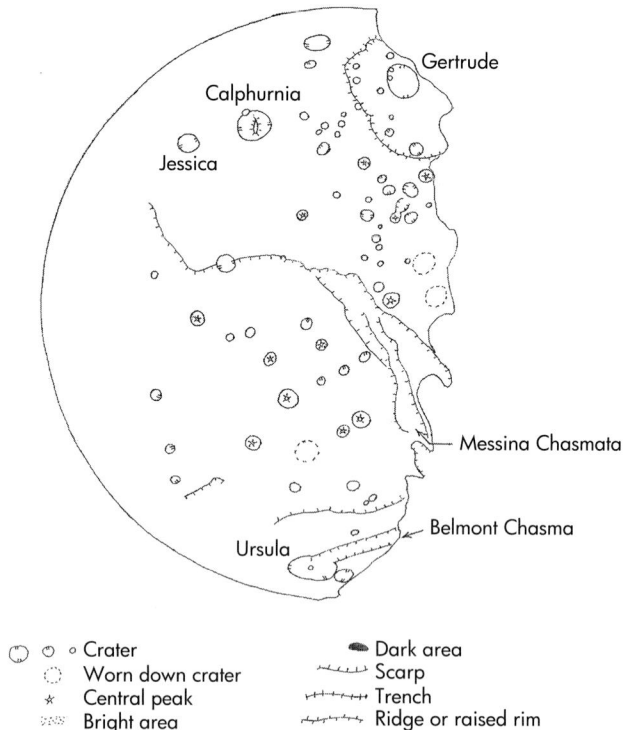

Figure 1.30. Map of part of Titania. The terminator is on the right side and is jagged due to shadows. (Credit: Richard W. Schmude, Jr.)

In addition to craters, Titania also has several faults, scarps and canyons. One canyon, Messina Chasmata, is up to 100 km across, over 1,500 km long and at least 1.0 km deep. As a comparison, the Grand Canyon is less than 600 km long. Several of the canyons on Titania formed over a long period and in several stages. We know that the canyons are relatively young because they have low crater densities. Titania has several scarps, or cliff-like features cutting across its surface, and one of them is four kilometers high.

A possible scenario for Titania's geologic history is that it was first bombarded by meteoroids of up to several tens of kilometers across. This bombardment created lots of craters with diameters exceeding 100 km. One or more episodes of resurfacing then buried the craters, including the larger ones. The molten material responsible for the resurfacing was at least a couple of kilometers thick since craters over 100 km across were buried. Once the resurfacing phase ended, the surface solidified and the meteoroid bombardment continued but at a slower rate than previously. At some point, Titania expanded a little causing the canyons and faults to form. Finally, additional meteoroids struck that moon creating craters with visible ejecta.

Spectroscopic studies show that water ice and carbon dioxide ice cover parts of Titania's surface. We also know that the water ice is in a crystalline state, which means that the ice molecules are arranged in a regular pattern instead of a random one. The spectroscopic signal of water is stronger on the leading side than on the following side; however, the reverse of this is true for carbon dioxide ice.

One group has examined the asymmetrical distribution of water ice and carbon dioxide ice on Uranus's large moons. They suggest that the asymmetry may be due to either charged particles or ring particles striking the surface of this moon and causing changes. Since Uranus rotates much faster than the regular satellites, the magnetic field and trapped particles collide with the trailing hemispheres of Ariel, Umbriel and Titania more often than their leading hemispheres. These trapped particles may affect the spectral signatures of water and carbon dioxide ice. Oberon lies far enough out that it may not be affected as much by trapped particles.

Brightness measurements made by Voyager 2 at a wavelength of 0.48 µm enabled astronomers to construct the phase curve of Titania. See Figure 1.31. The horizontal axis shows the solar phase angle of that moon with respect to the probe. (The solar phase angle is the angle between the Sun and the target measured from the observer's location, which in this case is Voyager 2.) The vertical axis shows the drop in brightness. At phase angles below ~2°, Titania brightens rapidly. The difference between the brightness extrapolated from phase angles exceeding 10° and the actual brightness at opposition is the opposition surge. The quantity F in Figure 1.31 is Titania's opposition surge and it equals ~0.3 magnitudes in 0.48 µm light.

Titania has a substantial opposition surge in near-infrared light (wavelength 2.2 µm). Oberon and Ariel also have large brightness surges at opposition in this light.

Astronomers know that Titania's surface is very porous because of its large opposition surge. This should not be surprising given the fact that Titania's gravity is very weak compared to Earth's gravity. Table 1.10 lists the weight that a 200 pound man would have on several moons along with their gravitational acceleration and escape speeds. As it turns out, a 200 pound man on Earth would weigh only 7.6 pounds on Titania. The low weight means that the soil is less compressed.

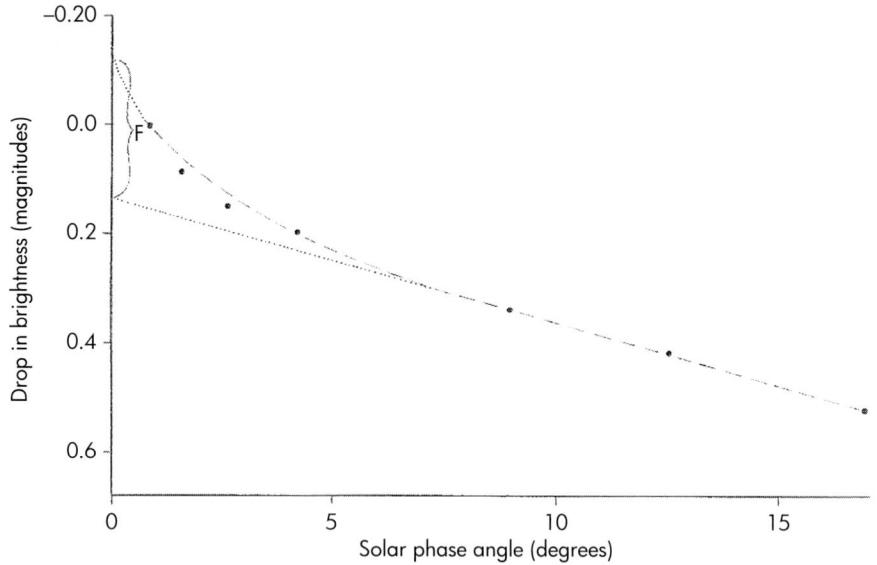

Figure 1.31. Phase curve of Titania. This graph shows the drop in brightness of Titania (vertical axis) at different solar phase angles (horizontal axis). All brightness drops are with respect to the brightness measured at a solar phase angle of 0.8°. The data are based on clear filter measurements made by Voyager 2. The opposition surge, F, is the difference between the brightness at a solar phase angle of 0.0° and the value extrapolated form phase angles exceeding 10°. (Credit: Richard W. Schmude, Jr.)

The albedo is the fraction of light that a surface reflects. The geometric albedo is defined in the Astronomical Almanac for the year 2007 as "the ratio of the illumination of the planet at zero phase angle to the illumination produced by a plane, absolutely white Lambert surface of the same radius and position as the planet." page E4. One problem with reporting the geometric albedo is getting measurements at a solar phase angle of exactly 0°. Even if measurements are made at a solar phase angle of 1° they will be off because many objects brighten by several percent or more as the solar phase angle drops from 1°–0°. The albedo of Titania's southern hemisphere based on the Voyager data is near 0.28. The albedo

Table 1.10. Surface gravity and escape speed of some of the moons of Uranus

Object	Gravitational acceleration at the surface (m/s^2)	Weight of a 200 pound man (pounds)	Escape speed (m/s)
Earth	9.8	200	11,200
Miranda	0.079	1.6	193
Ariel	0.25	5.2	541
Umbriel	0.25	5.1	542
Titania	0.37	7.6	766
Oberon	0.35	7.1	727
Puck	0.034	0.7	74
Cordelia	0.0096	0.2	21

is about 0.34 if one considers Earth-based data with light having a wavelength of 0.54 µm. Keep in mind that in 2008, both the northern and southern hemispheres of Titania will face Earth; hence, these values may not apply. In 2030, the northern hemispheres of the regular moons will face Earth. It will be interesting to compare at that time the albedos of Titania's northern and southern hemispheres.

The slope of the phase curve at a solar phase angle of 20°–60° is the solar phase angle coefficient; for Titania's southern hemisphere, it equals 0.02 magnitudes per degree. This value is similar to that of our moon, but is much greater than what it is for Uranus. A smooth icy surface with few shadows will have a lower solar phase angle coefficient than one that is fluffy with lots of holes and shadows. The value of Titania's solar phase angle coefficient is consistent with it having a fluffy surface.

Titania can brighten by up to about 0.8 magnitudes during an apparition. (An apparition lasts about 370 days for Uranus, and it is the length of time between when Uranus has the same right ascension as the Sun to when it again has this characteristic.) Much of the brightness change is due to the opposition surge. One group reports a solar phase angle coefficient of 0.102 ± 0.021 magnitude/degree for Titania based on V filter measurements made in 1982 and 1983 at solar phase angles of between 0.2° and 3°.

Titania's surface reflects about one-fourth of the visible light that falls on it and absorbs the remaining amount. Maximum surface temperatures of about 87 K (–303° F) probably occur on that moon during the summer. Very dark areas on Titania may even get a bit warmer because they absorb more light. During the long winters the temperatures may drop to around 30 K. Titania reflects about the same fraction of blue, green and red light falling on it; however, since the Sun gives off more red than blue light, Titania reflects more red than blue light. One astronomer has analyzed how Titania reflects different colors of light and concludes that this moon has a yellow color. One will only be able to see this color though through a two-meter telescope since this moon is so faint. See Chapter 4.

We are not sure if Titania has a thin atmosphere or not; furthermore, the presence or absence of an atmosphere on it can place constraints on its surface composition. Voyager 2 did not detect an atmosphere around Titania. Antonio Cidadao recorded an occultation of the star HIP 106829 by Titania on September 8, 2001. His data shows that if Titania has an atmosphere then it is extremely thin. Antonio's experiment is described in more detail in Chapter 6. Methane ice has a melting point of 91 K; hence, sunlight alone will not likely melt methane on Titania. Methane ice, however, has a vapor pressure of 0.01 bar at 76 K. Therefore, any methane ice in Titania's southern hemisphere would release enough gas to have been detected. Consequently, large amounts of pure methane ice are not present; however, methane may be dissolved in water ice or other ices.

Due to the lack of a significant atmosphere, Titania's surface is undoubtedly bombarded by meteoroids of all sizes. Its surface is undoubtedly covered with lots of craters having diameters of less than a few meters.

Oberon

Close-up images of Oberon have enabled astronomers to map its surface and to measure its polar and equatorial diameters. A Voyager 2 image of much of Oberon's southern hemisphere is shown in Figure 1.32. Each pixel in this image is six kilometers across, which makes features less than about twelve kilometers

Figure 1.32. Voyager 2 image of Oberon made on Jan. 24, 1986 at a distance of 660,000 km. (Credit: Courtesy NASA/JPL-Caltech.)

difficult to identify. In fact, many of the canyons on Miranda would not be visible at the resolution of Figure 1.32. Oberon's surface is covered with craters, with some believing that its surface is saturated with craters. This means that the number of craters being destroyed equals the number being formed. The density of craters with diameters ≥ 30 km on Oberon is over twice that found in the cratered terrain on Titania. Oberon's polar diameter is within four kilometers of its equatorial diameter and, hence, its ellipticity is <0.003. This is consistent with its slow rotation.

Figure 1.33 shows a map of part of Oberon. Simple and complex craters cover Oberon. Shadow studies reveal that the depth-to-diameter ratio of large simple craters is 0.1, which is about half that of similar sized craters on Earth's Moon. Essentially, the craters on Oberon are shallow. The shallow craters may be due to the icy nature of its surface. One prominent complex crater is Hamlet, which has a diameter of ~200 km. This crater has a central peak and has dark material on its floor with an albedo of 0.14. This material has the lowest albedo of any area imaged on Uranus's regular moons and is believed to have come up from Oberon's interior. A second, prominent crater, Othello, is ~90 km across. Three bright rays extend from this crater to the south pole. These rays are at least 300 km long and may be similar to those found on our moon. Mommur Chasma is a large canyon that lies near Oberon's equator. This feature is up to 40 km wide, 400 km long and is about three kilometers deep. Oberon also has a mountain on its limb that is 11 km high and 45 km wide at the base. For a comparison, the top of Mount Everest, the highest point on Earth, is less than nine kilometers above sea level. Oberon's limb mountain may be the central peak of a 400–500 km basin.

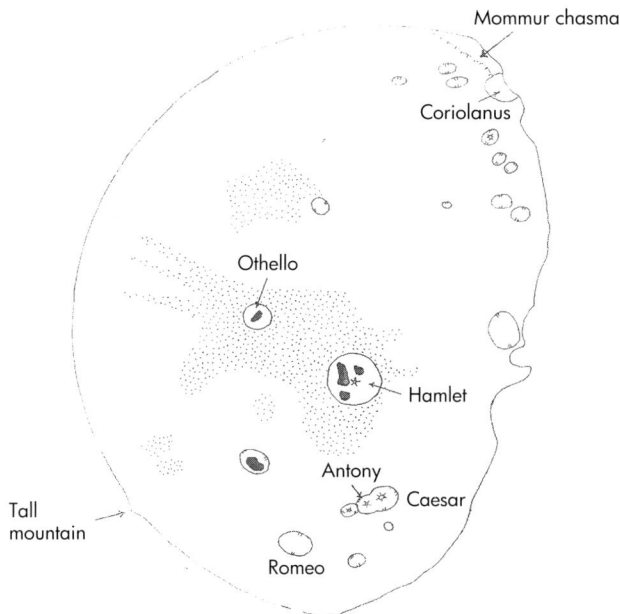

Figure 1.33. Map of part of Oberon. Symbols are the same as in Figure 1.30. (Credit: Richard W. Schmude, Jr.)

Oberon probably had significant geological activity in its past. Oberon's large canyons are older than those on Titania, which suggests that Oberon's geological activity ceased much sooner than Titania's activity. One possible scenario for Oberon's history is that soon after it formed and solidified, it underwent a heavy bombardment by meteoroids having diameters of up to several tens of kilometers. The bombardment created craters of up to a few hundred kilometers in size. Early in Oberon's history, a limited amount of resurfacing took place which buried some, but not all of the large craters. The surface continued to be bombarded by meteoroids but at a slower rate. Oberon then expanded creating large canyons. Meteoroids continued to rain down on Oberon for an extended time.

Oberon's surface has water ice on it. This ice is in the crystalline state, which is the same as the ice on Titania. Unlike Titania though, ice appears to be more abundant on Oberon's following hemisphere. A second difference between Oberon and Titania's surface is that astronomers have not detected carbon dioxide ice on Oberon.

Voyager 2 yielded brightness data of Oberon's southern hemisphere at several wavelengths. No measurements at low phase angles, however, were made. Oberon is a little darker than Titania. Its surface displays a wide range of albedos and colors. In fact, Oberon has a wider range of color than any of the other regular moons of Uranus. Oberon reflects more yellow than blue light. Like Titania, Oberon has a substantial opposition surge. This is consistent with that moon having porous soil like Titania.

Since Oberon has the same seasons as Titania, it probably has similar temperatures. The peak summertime surface temperature may reach 85–90 K and possibly a little more in the very dark areas. The winter temperatures probably drop to around 30 K.

Umbriel

Figure 1.34 shows an image of Umbriel's southern hemisphere. One can identify features as small as 10 km in this image. Like Oberon, Umbriel is covered with craters including large ones. The crater density (diameters $\geq$30 km) is over three times that of the cratered terrain on Titania. The part of Umbriel shown in Figure 1.34 is as old as Oberon and is older than Titania. Limb profiles reveal topographic variations of up to 10 km. This is more than what is found on Ariel and Titania. Umbriel has a nearly spherical shape.

Figure 1.35 shows a map of part of Umbriel. The $\sim$150 km crater Wunda is the most distinct feature. Wunda is covered with bright material that reflects more light than the surrounding areas and it has an albedo near 0.5. A second large crater, Vuver has a diameter of $\sim$100 km. This crater has a central peak that is partially covered with bright material that is probably the same as what is covering Wunda. Umbriel has several scarps. One of them is shown on the map. One scarp is over 300 km long.

One possible scenario of Umbriel's geological history is that after formation and surface solidification, it was blasted with meteoroids of varying sizes. Some of these meteoroids had diameters exceeding 20 km and created craters over 200 km across. Parts of Umbriel underwent resurfacing events. This resurfacing covered some of the craters but left others intact. Meteoroids continued to collide with Umbriel creating additional craters. At some point, that moon underwent expansion and canyons formed. The expansion was not as extensive as on Titania or Ariel. Later on, small amounts of bright material erupted on the floor of the crater Wunda. Craters with ejecta deposits formed later and the ejecta covered older craters.

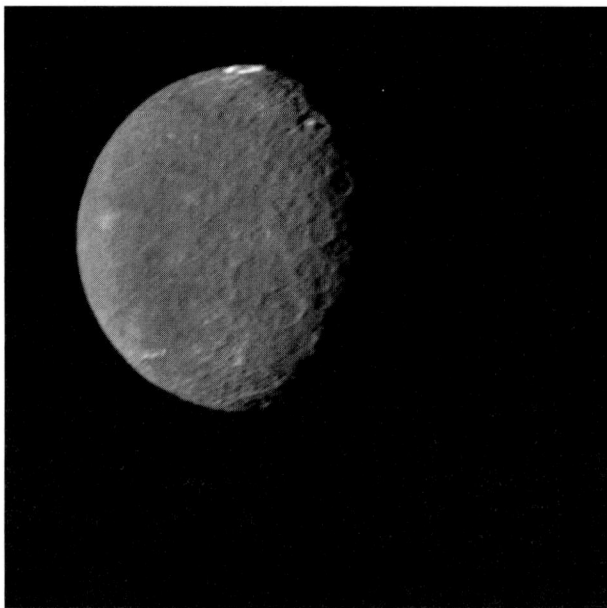

Figure 1.34. Voyager 2 image of Umbriel made on Jan. 24, 1986. (Credit: Courtesy NASA/JPL-Caltech.)

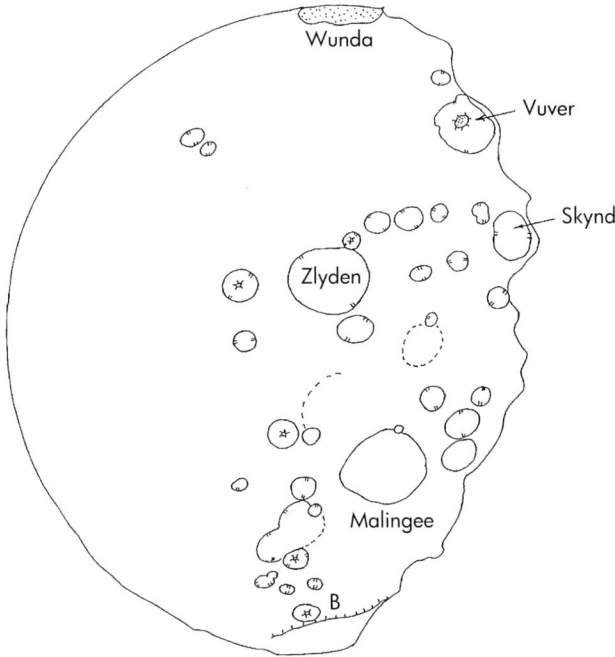

Figure 1.35. Map of part of Umbriel. Symbols are the same as in Figure 1.30. (Credit: Richard W. Schmude, Jr.)

Spectroscopic studies reveal that both water ice (in crystalline form) and carbon dioxide ice are on Umbriel. The spectroscopic signal of water is stronger on the leading hemisphere than on the following hemisphere. The water ice signal on Umbriel is weaker than on the other four large moons. This may be due to the greater abundance of dark material on that moon, which would also explain its low albedo. The carbon dioxide ice signal is stronger on the trailing than the leading hemisphere of Umbriel.

All parts of Umbriel's southern hemisphere have almost the same albedo. This is different from the other regular moons. Umbriel is also darker than the other four large moons. Geometric albedos for Umbriel's southern hemisphere are 0.17, 0.19 and 0.17 for ultraviolet, blue-green and near-infrared light respectively. These values are ~50% lower than those for Titania. One astronomer states that Umbriel has a reddish color.

Ariel

Close-up images have allowed astronomers to map several features on Ariel. Figure 1.36 shows a Voyager 2 image of the southern hemisphere of that moon. One can identify features as small as two kilometers in this image. Most of the image is on the side facing Uranus. Since Ariel was closer to Voyager 2 as that probe passed Uranus, the resolution in the figure was better than for Titania, Oberon or Umbriel. Many simple craters and a few complex ones are visible. A few of these have bright ejecta. Some of the ejecta blankets have albedos of 0.55, which

Figure 1.36. Voyager 2 image of Ariel made on Jan. 24, 1986, at a distance of 130,000 km. (Credit: Courtesy NASA/JPL-Caltech.)

is among the brightest areas on any of the large moons of Uranus. A few 8–12 km craters are very shallow. This may be due to recent geological activity on that moon or the icy nature of the surface. The transition size between simple craters and complex ones is around 15 km for Ariel. Unlike Oberon and Umbriel, Ariel has very few large craters. It has radii of 581, 578 and 578 km. The longest dimension points towards Uranus. This shape is consistent with its rotation speed and with a synchronous orbit.

Figure 1.37 shows a map of some of the major geological features on Ariel. One large crater, Yangoor, is about 80 km across and it has a central peak. Part of this crater is covered up with younger material. A second complex crater, Melusine, is surrounded by bright terrain, which is believed to be crater ejecta. An extensive canyon system lies just south of Melusine. The canyons on Ariel are 10–100 km wide and are up to 4 km deep. One canyon, Kachina Chasmata, is over 500 km long.

Crater density studies reveal that Ariel's surface has a wide range of ages. Many parts of that moon have crater densities much lower than in the cratered terrain on Titania. The crater density and surface features have given astronomers hints to its past. A possible scenario for Ariel's geological history is that after it formed and solidified, its surface became saturated with craters as a result of the same heavy bombardment that took place on the other moons. Early in Ariel's history, it was resurfaced. This resurfacing buried all of the older craters. That moon then expanded, creating several faults, scarps and canyons. The surface continued to be bombarded by meteoroids but at a slower rate than before the resurfacing; minor episodes of faulting continued. Later on, large blocks of cratered terrain broke off and some minor faulting and resurfacing continued.

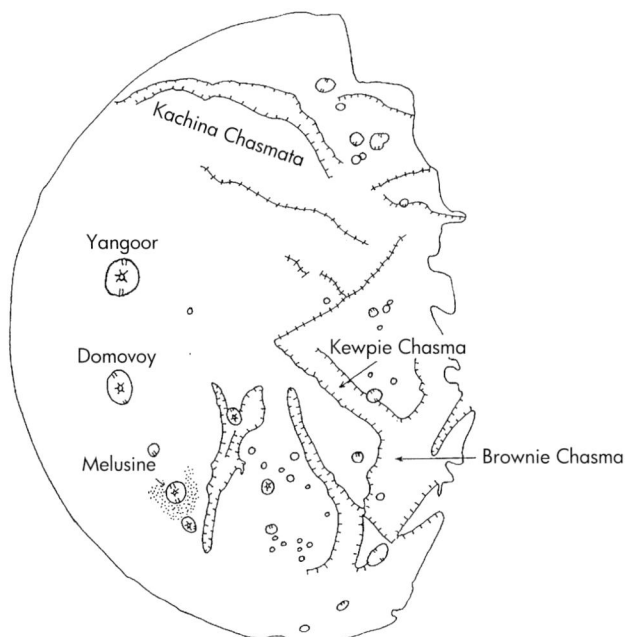

Figure 1.37. Map of part of Ariel. Symbols are the same as in Figure 1.30. (Credit: Richard W. Schmude, Jr.)

Both water ice and carbon dioxide ice are on Ariel. The water ice is in a crystalline state. Like Titania, the spectroscopic signal of water is strongest on the leading hemisphere. The carbon dioxide ice appears to be more abundant on the trailing hemisphere. The spectroscopic signals for carbon dioxide ice are stronger on Ariel than on Umbriel and Titania.

Like the other regular moons, Voyager 2 yielded data on Ariel's brightness at several wavelengths of visible light. This probe, however, did not collect data at low solar phase angles, but when Voyager 2 data is combined with Earth-based data it is apparent that Ariel has a large opposition surge like Titania. Ariel's surface has a high porosity. We know this from Voyager 2 thermal data and Ariel's large opposition surge. This should not be surprising due to the low gravity on that moon. A 200 pound man on Earth would only weigh 5.2 pounds on Ariel. One astronomer describes Ariel as having a yellow-orange color. Its southern hemisphere has a higher albedo than the southern hemispheres of the other four regular moons.

Miranda

Miranda is perhaps the most bizarre moon in the Solar System. Figure 1.38 shows the southern hemisphere of that moon. One can identify features smaller than one kilometer in this figure. Miranda has several cliffs, faults, ridges and canyons on its surface. Figure 1.39 shows a map of part of Miranda. The surface of Miranda contains four types of terrain, which are the corona, heavily cratered terrain, lightly cratered terrain and terrain dominated by faults, scarps and canyons.

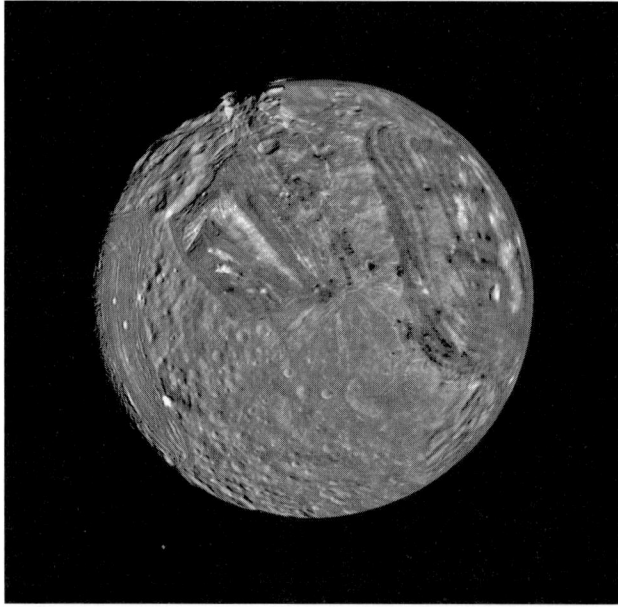

Figure 1.38. Voyager 2 image of Miranda made on Jan. 24, 1986. This image is a mosaic of several close-up images. (Credit: Courtesy NASA/JPL-Caltech.)

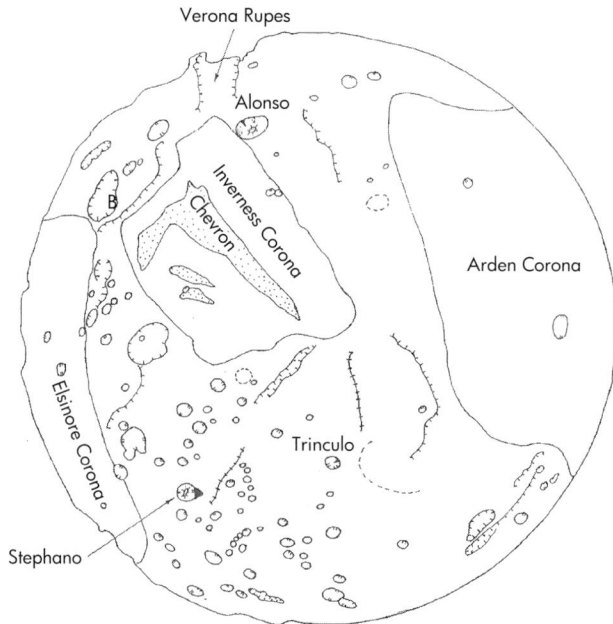

Figure 1.39. Map of part of Miranda. Symbols are the same as in Figure 1.30. (Credit: Richard W. Schmude, Jr.)

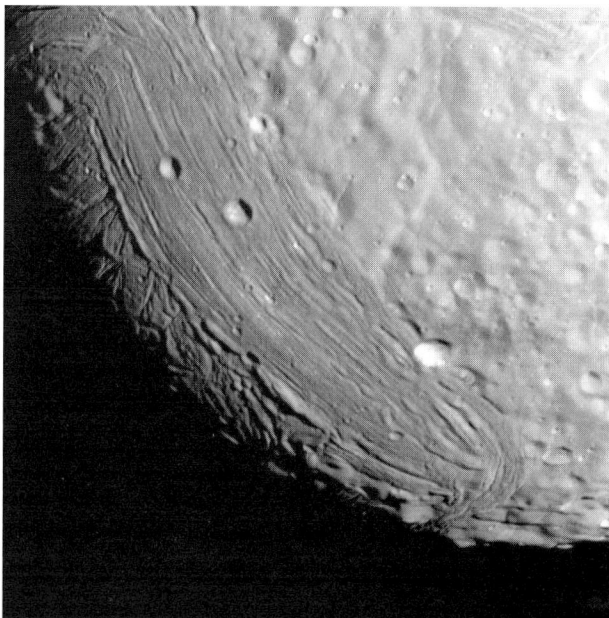

Figure 1.40. Voyager 2 image of Elsinore Corona and surrounding areas on Miranda. The image covers an area of 200 by 230 km. (Credit: Courtesy NASA/JPL-Caltech.)

Elsinore Corona, Inverness Corona and Arden Corona are much different than the surrounding terrain. These three areas all have low crater densities, which show that they are young. Inverness Corona contains lots of ridges. It has a bright area that looks like either a chevron or the number 7 upside down. I call it a chevron. The chevron is up to twice as bright as the surrounding terrain. The central regions of Elsinore Corona have higher elevations than the surrounding cratered terrain. This corona contains several ridges, which are ~500 meters high. See Figure 1.40. Arden Corona lacks the ridges that are found on the other two corona. A deep canyon surrounds this corona.

How did Elsinore, Inverness and Arden Corona form? Astronomers have proposed several hypotheses to answer this question. One of these is that a violent collision between Miranda and a giant object occurred which broke this moon into many pieces. Later on these pieces came back together forming what we see as Miranda. One group proposed a second hypothesis which was that Arden Corona is the remains of a giant impact crater.

In addition to corona, Miranda also has a wide variety of craters. Crater B (diameter ~30 km) has a central mound. This crater is probably a worn down complex crater. There are no other obvious complex craters in Figure 1.38. A few craters like the 7 km crater Trinculo have bright bands of material on their slopes, while others like the ~10 km crater Stephano have dark ejecta blankets surrounding them. One of the most distinct craters on Miranda is the ~25 km crater Alonso. See Figures 1.39 and 1.41. This crater appears to have some bright and light gray material on its inner walls.

Canyons are almost as common as craters on Miranda. One scientist has identified four types of canyons, which are called types 1, 2, 3 and 4. The type 1

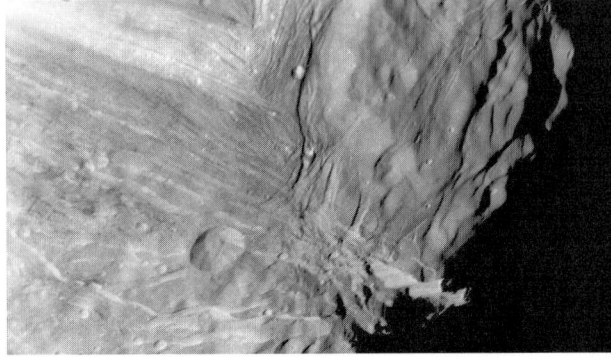

Figure 1.41. Voyager 2 image of the 25 km crater Alonso (a little below and left of the center) Note the bright streaks on the inner walls of this craters. The huge cliff is part of the large canyon Verona Rupes. This image shows an area 250 km across. (Credit: Courtesy NASA/JPL-Caltech.)

canyon is 1–2 km wide and is up to 40 km long. The type 2 canyon is a little larger, being up to 20 km wide. Type 3 and 4 canyons are around 35 and up to 80 km wide. These canyons can be over 100 km long. Verona Rupes is an example of a type 4 canyon. The canyon walls can be quite steep. In one case, a cliff higher than Mount Everest lies near Miranda's equator in the canyon Verona Rupes. See Figure 1.41.

Miranda is not quite round. It has dimensions of 480, 468 and 466 km. The largest one, 480 km, points towards Uranus. Since Miranda is smaller than the other four regular moons, it has a weaker gravitational field, and as a result, it is not able to pull itself into as round of a shape as the larger moons.

Miranda's surface is similar to that of the other large moons. Water ice, in crystalline form, covers much of its surface. This ice is mixed with darker material that probably contains rock and carbon-rich material. Miranda's surface is very porous. This should not be surprising, considering that a 200 pound man on Earth would only weigh 1.6 pounds on that moon.

Miranda reflects a little more blue light than orange and red light; hence, it has a bluish color. This is different from Uranus's other regular moons. This difference may be due to a greater percentage of ice on Miranda. Like Uranus, Miranda reflects less near-infrared light than visible light.

We are not sure what Miranda's interior is like, but this moon's mass and size yield clues. Miranda has a density that is 20–30% lower than that of the other regular moons; hence, it probably contains a greater percentage of ices than the others. Miranda's interior is probably quite cool compared to the larger moons for two reasons: its low density and small size. Let me explain this. This moon probably has a very low percentage of heavy radioactive elements because of its low density; hence, it probably has very little internal heat that comes from these elements. Furthermore, most of the internal heat that Miranda had during its formation phase would have escaped due to its small size.

Voyager 2 measured a maximum summertime temperature of 86 K on Miranda. Some of the dark areas in Inverness Corona may warm up to 90 K during the summer since dark areas absorb more sunlight. Temperatures may drop to around 30 K during the long winter.

Collision Fragments

There are currently 13 known moons in the collision fragment category. These are important because many of them have an intermediate position between the inner rings and the large moons. They may yield information on how the rings formed. These moons also exert gravitational tugs on the rings and on each other. One of the collision fragment moons, Mab, appears to be a source of particles for one of the rings. Since the moons in this group are so close to Uranus, they move very fast. Figure 1.42 shows an image of Uranus, its rings and three inner moons. The inner moons appear as several dots in this image because of their high speeds.

One astronomer re-analyzed Voyager 2 images in the late 1990s and was able to determine approximate diameters, albedos and sizes for these moons. Most of them have irregular shapes with their longest axes pointed towards Uranus. This orientation is consistent with synchronous rotation. They all have low albedos, and there is some evidence that Puck has a higher albedo than the others. We have no information about topographic features on any of them except for Puck.

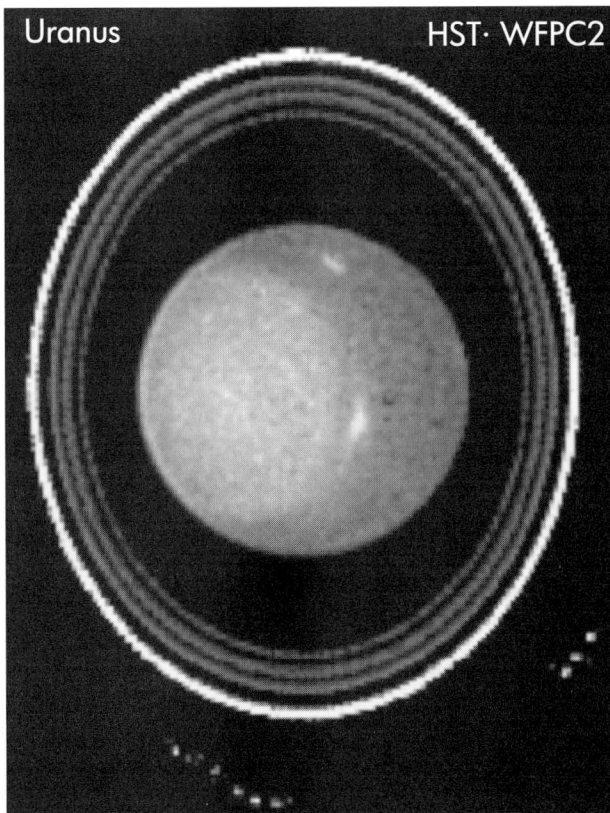

Figure 1.42. Hubble Space Telescope image of Uranus, its rings and three of its small inner moons. Since the inner moons move so fast they each appear as a string of dots. Several of the rings are visible along with two bright clouds. The bright oval on Uranus's disc is high altitude haze that lies over the planet's south polar region. (Credit: Courtesy NASA/JPL-Caltech.)

Why do the inner moons have such low albedos? One possibility is that due to their small sizes they never underwent differentiation. As a result, not as much bright water ice made its way to the surface. A second explanation for the low albedos of the collision fragment moons is radiation darkening. Scientists have shown that fast moving sub-atomic particles, like what are in Uranus's magneto-sphere, can break chemical bonds in methane ice causing hydrogen to escape and leaving behind a dark carbon residue. Since the collision fragment moons lie close to Uranus, they may receive a more intense bombardment of sub-atomic particles than the regular satellites.

The orbits of the inner moons may undergo small changes from year to year. One group reports that Mab underwent a $\sim 1°$ change in its orbit around 2004. Resonances with small, undiscovered moons may be causing small changes in position.

The densities of the inner moons are assumed to equal 1.5 g/cm^3, which is near the average density of the five large moons. This writer computed volumes by assuming a triaxial ellipsoid shape with the third unknown dimension equal to the smaller of the two measured dimensions. With these volumes and the assumed density, approximate masses were computed and are listed in Table 1.8. A few individual collision fragment moons are discussed.

Puck

Figure 1.43 shows an image of Puck. One would not be able to identify features smaller than about 10 km in this image due to its limited resolution. One large crater, Bogle, which is almost 50 km across, is near the right edge. This crater is

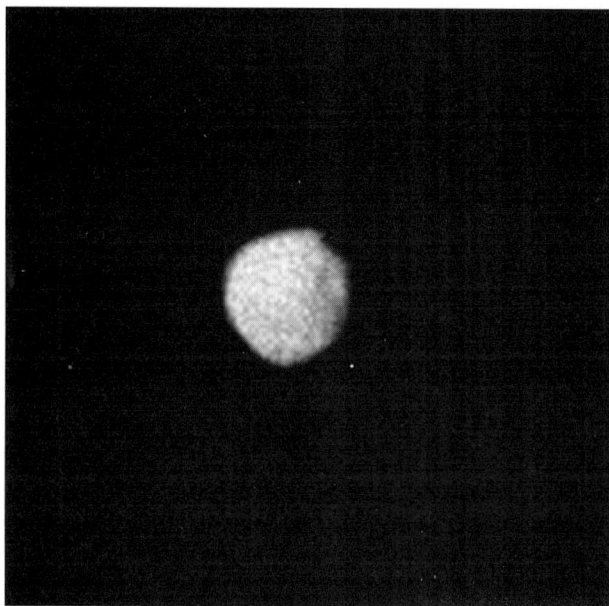

Figure 1.43. Voyager 2 image of Puck. (Credit: NASA and the National Space Science Data Center.)

almost one-third the size of Puck. The event that created Bogle probably came close to breaking this moon apart. Two smaller craters, Butz and Lob, are about 25 km across and are at the bottom of the image.

Puck's shape and orientation are consistent with it having a synchronous orbit. Two of the three radii are 82 and 77 km; we are not sure of the dimension perpendicular to the orbital plane. Puck is one of only two moons in the collision fragment group whose period of revolution is longer than Uranus's rotational period.

The portion of Puck's surface imaged by Voyager 2 has a nearly uniform albedo. Puck is quite dark and it reflects only about 10% of the visible light falling on it at opposition. Its albedo is higher in near-infrared light with a wavelength of 0.91 μm than in green light, and its B–V color index is similar to that of the four largest moons of Uranus. According to one astronomer, Puck has a brownish color. Infrared data is consistent with the presence of water ice. When this moon's solar phase angle is 40°, its solar phase angle coefficient is near 0.025 magnitude/degree. This means that it gets 0.025 magnitudes (or 2.5%) brighter when its solar phase angle drops from say 40° to 39°. From Earth, we only see Puck at solar phase angles below 3°. It probably has a large opposition surge and, hence, will brighten much more than 0.025 magnitudes/degree as seen from Earth.

Portia

Voyager 2 was unable to get a close-up image of this moon, but we do know its approximate size; furthermore we know its brightness in different colors of light, and this has given us some information about this moon's characteristics. Its brightness and size are consistent with it reflecting about 10% of the visible light falling on it at opposition. Portia is reported to have a brownish color, and it has a similar B–V color index as Puck. Near-infrared measurements of Portia are similar to those of Puck and are consistent with the presence of water ice and possibly carbon-based material on its surface. Portia has a somewhat irregular shape. Its longest dimension is approximately 1.2 times its short dimension. As a result, its brightness changes by up to ~0.2 magnitudes as it moves around Uranus. Portia's longest dimension is about the same as Puck's and, hence, this moon appears to be a close second in size to Puck.

Juliet and Belinda

These two moons have irregular shapes. Their long dimension is about twice their short dimension and, hence, their shapes are close to that of a potato or a rugby ball. As a result, their brightness changes as they move around Uranus. Table 1.9 lists approximate brightness values for these moons. Each moon reflects about 10% of the visible light falling on it at opposition. These moons reflect also around 8–10% of the near-infrared light falling on them at opposition.

Cordelia and Ophelia

Cordelia and Ophelia both lie near the Epsilon ring and are believed to be responsible for constraining this ring. These moons have similar shapes; their longest dimension is about 1.4 times their shorter dimension. Because of this, their brightness change as they move around Uranus.

Captured Objects

Astronomers discovered nine small moons lying outside the orbit of Oberon (Francisco, Caliban, Stephano, Trinculo, Sycorax, Margaret, Prospero, Setebos, and Ferdinand) between 1997 and 2003. Both Hubble Space Telescope images and those from ground based telescopes have yielded what we know about these moons. Because of the limited resolutions of these telescopes, we know very little about their shapes and sizes. All of these moons, except for Margaret, move in a retrograde orbit. All nine outer moons move in eccentric and highly inclined orbits, which is consistent with them being captured objects. Due to their small sizes, they probably have irregular shapes, but we are not certain of this. If one assumes that the nine outer moons all have a spherical shape and have the same albedo as Puck, one can compute approximate diameters from brightness measurements. This has been done, and based on one study, the computed size distribution is inconsistent with these moons coming from one collision. The assumption of constant albedos for these moons, therefore, may not be valid. Alternatively, as suggested by one group, Uranus's nine outer moons may represent more than one set of collision fragments.

Data in Tables 1.8 and 1.9 are based on brightness data along with assumed shapes, densities and albedos of the moons. The radii in Table 1.8 for all of the moons beyond Oberon were computed using the magnitudes in Table 1.9 and an assumed albedo of 0.15. This is the average albedo of 22 Trans-Neptune Objects (TNOs) according to a recent study. The mass values were computed from the radii in Table 1.8 along with assumed density values of $1.5\,g/cm^3$.

There may be additional moons orbiting Uranus. The region where stable satellite orbits occur is called the planet's Hill sphere. For Uranus, the Hill sphere has a radius of about 70 million km, which is much greater than the mean distance between Uranus and its outermost known moon, which is ∼21 million km. There may be more small moons lying beyond Ferdinand.

All of the outer moons lie at average distances of at least seven times that of the outermost regular satellite, Oberon. Gravity decreases as the square of the distance, and so these moons feel less than 2% of the gravity that Oberon feels from Uranus. As a result, they may not undergo synchronous rotation; furthermore, tidal forces from Uranus are small for moons beyond Oberon.

These nine moons lie generally outside of Uranus's magnetosphere and, hence, they do not receive the bombardment of charged particles that the other moons receive. Their surfaces, therefore, may be much different than those on the inner moons. I will discuss the two brightest moons in this group – Sycorax and Caliban.

Sycorax

Sycorax is the brightest of the collision fragment moons in visible and near-infrared light. In fact, one Canadian amateur astronomer imaged Sycorax in 1999. The limiting magnitude of this individual's equipment is around magnitude 21 and, hence, his image is consistent with the magnitude values listed in Table 1.9.

Sycorax reflects more red than green and blue light. Its B–V, V–R and R–I color indexes are 0.7–1.0, 0.5 and 0.5–0.6, respectively, which are close to the values for Pluto and other trans-Neptune objects (TNOs). This moon is also redder than the Sun. Sycorax reflects about the same percentage of light in the J, H and K bands. This is, however, different from the behavior of TNOs. The spectrum of Sycorax is also different than for Caliban. Some astronomers believe that this is evidence that these two moons came from two different parent bodies.

Sycorax may rotate once every 4 hours; however more measurements are needed for confirmation. Since Sycorax lies so far from Uranus, it probably does not have a synchronous rotation.

One Russian astronomer carried out extensive calculations on the orbit of Sycorax. Based on these calculations, this individual reports that its orbital inclination has changed by $\sim 7°$ and that its orbital eccentricity has changed by $\sim 10\%$ over the last 250 years. Much of this change is a result of the large distance between Sycorax and Uranus.

Caliban

Caliban is the second-brightest of the nine moons beyond Oberon. Its orbit is just over half the size of Sycorax's orbit. This moon has a color similar to that of Sycorax's except for light having a wavelength of 0.7 μm. Apparently, Caliban absorbs light with this wavelength. One group of astronomers points out that this is evidence that liquid water may have modified its surface. This moon may rotate once every ~ 3 hours; however, more measurements are needed for confirmation.

The Neptune System

Introduction

Before 1975, Uranus and Neptune were thought to be very similar to each other. After all, they are gaseous planets and have similar sizes, masses and colors. The problem with this thinking, though, was that we did not know much about either planet. We had little information on their meteorology, atmospheric composition and clouds. This began to change in 1976, when astronomers discovered that Neptune brightened by at least a factor of four in near-infrared (near-IR) light; furthermore, this brightness changed over several hours due to its rotation. This rise and fall in brightness was the first hint that things were different on Neptune. Over the next 30 years, our knowledge of Neptune has increased to the point where a more informed discussion of it can be presented.

This chapter is divided into eight sections. The first four summarize Neptune's Atmosphere, Interior, Magnetic Environment and Rings and Arcs. The last four summarize Neptune's satellites especially Triton, followed by Collision Fragments, Captured Objects and Trojan Asteroids. Table 2.1 lists a few characteristics of Neptune.

Atmosphere

In many ways, Neptune's atmosphere is similar to Uranus' atmosphere. It possesses clouds and undergoes brightness changes which appear to match the seasons. Like Uranus, Neptune does not have a visible surface; hence, altitudes are given with respect to the 1.0 bar level or in terms of the local atmospheric pressure. Like Earth, Neptune's axis is tilted and, as a result, its atmosphere experiences seasons. Our view of Neptune's atmosphere changes during its 165-year trip around the Sun. Hence, it will take over a century of observations to measure fully seasonal changes in Neptune's atmosphere, and this must be kept in mind when attempting to interpret recent data. I will start by discussing the upper, middle and lower layers of Neptune's atmosphere. This will be followed by discussions of clouds, brightness changes and other characteristics of Neptune's atmosphere.

Upper Atmosphere

Figure 2.1 shows a cross-section view of Neptune's upper atmosphere. As with Uranus, Neptune has an ionosphere which lies in the same area as the thermosphere. Above the thermosphere lies the exosphere.

R.W. Schmude, Jr., *Uranus, Neptune, and Pluto and How to Observe Them*,
DOI: 10.1007/978-0-387-76602-7_2, © Springer Science+Business Media, LLC 2008

Table 2.1. Characteristics of Neptune

Characteristic	Value
Equatorial Radius	$24{,}764 \pm 20$ km (1 bar level)
Polar Radius	$24{,}340 \pm 30$ km (1 bar level)
Surface Area	7.62×10^9 km^2 (1 bar level)
Mass	1.025×10^{26} kg
Density	1.640 g/cm^3
Period of Rotation (interior)	16.108 hours
Period of Revolution	165 years
Inclination	28.3°
Average Distance from the Sun	30.06 au
Orbital Inclination	1.8°
Orbital Eccentricity	0.01
Ellipticity	0.017 ± 0.001
Magnetic Field Strength	0.1 to 1 gauss
Vo	7.76
V(1,0)	-6.95 [a]
B–V	0.42 [a]
V–R	-0.33 [a]
y(1,0)	-6.92 [b]
b–y	0.14 [b]
Solar Phase Angle Coefficient (V filter)	0.0015 ± 0.004 [c]

[a] Seasonally averaged values.
[b] Average of 1983–1987 results, which is in between Neptune's equinox and solstice dates.
[c] Average value of John Westlfall's 2001 and 2002 measurements.

Our knowledge of the ionosphere is based on Voyager 2 occultation data which was taken in August 1989. At that time, the sunspot number was ~150, which is near the maximum value of around 200. Earth's ionosphere tends to gain more charged particles during sunspot maximum, and Neptune's ionosphere may follow a similar trend. The Voyager data was also collected near Neptune's morning terminator. Earth's ionosphere changes during the day and night, and this is probably the case with Neptune. Neptune's ionosphere begins at an altitude of ~500 km where the pressure is ~1 μbar. At this level, there are ~100 electrons/cm^3. The electron density rises to ~1500 electrons/cm^3 at an altitude of 1,400 km. This is more than a million times lower than the neutral gas density. In spite of this, the ionosphere contributes a significant amount of thermal energy to the neutral gas layer. During 1989, our ionosphere degraded signals from Voyager 2 and, as a result, we are not sure of the upper boundary of Neptune's ionosphere, and there is a chance that Voyager 2 passed through it. Voyager 2 detected the ion N^+ near its closest approach.

Neptune's ionosphere is made up mostly of electrons and protons (H^+). Minor amounts of the ions H_2^+, He^+ and N^+ are also present. This layer forms as a result of ultraviolet radiation from the Sun ionizing neutral atoms. Important ionospheric reactions on Neptune probably include:

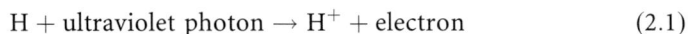

$$H + \text{ultraviolet photon} \rightarrow H^+ + \text{electron} \tag{2.1}$$

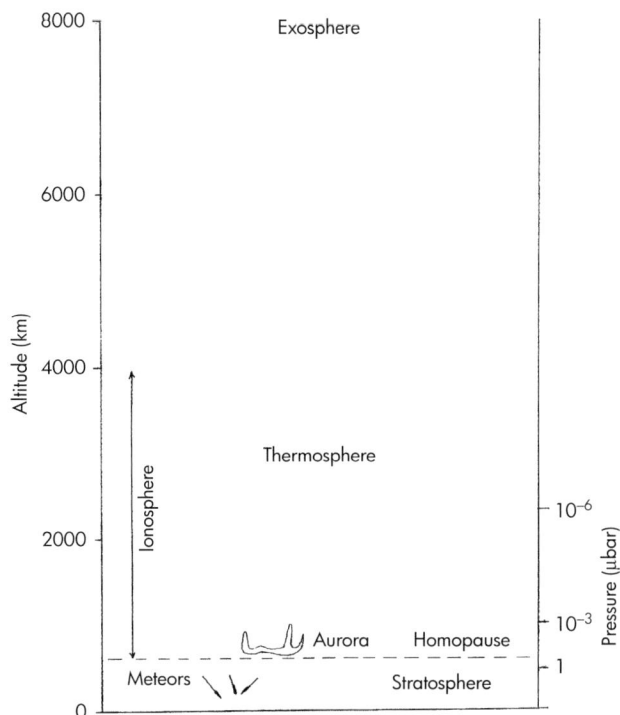

Figure 2.1. A diagram of Neptune's upper atmosphere. The altitudes are with respect to the 1.0 bar level. Different levels of pressure are shown at the right. (Credit: Richard W. Schmude, Jr.)

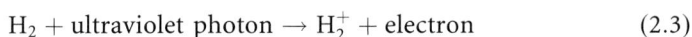

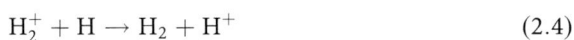

$$H + electron \rightarrow H^+ + 2 \ electrons \qquad (2.2)$$

$$H_2 + ultraviolet \ photon \rightarrow H_2^+ + electron \qquad (2.3)$$

$$H_2^+ + H \rightarrow H_2 + H^+ \qquad (2.4)$$

Additional reactions that may be important are:

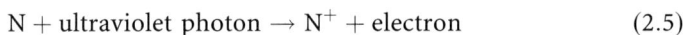

$$N + ultraviolet \ photon \rightarrow N^+ + electron \qquad (2.5)$$

$$N + electron \rightarrow N^+ + 2 \ electrons \qquad (2.6)$$

The thermosphere lies at the same level as the ionosphere and it has its own characteristics. It extends from about 500 km up to a few thousand kilometers above the 1.0 bar level. The temperature starts out at ∼160 K but rises to at least 550 K at an altitude of 2,000 km. Small amounts of hydrocarbons and nitrogen atoms are present. As in the case of Uranus, the fraction of atomic hydrogen rises with increasing altitude throughout the thermosphere.

There is some uncertainty as to the location of the homopause, and it may be near the 1.0 μbar level. The high altitude of the homopause compared to that of Uranus (20 μbar) may be due to stronger convection currents in Neptune's atmosphere.

The exosphere lies above the thermosphere. One difference between Uranus and Neptune is that nitrogen from Neptune's moon Triton falls back onto

Neptune. As a result, nitrogen enters Neptune's exosphere and, at the same time, hydrogen escapes.

Does Neptune have aurorae? We are not sure of the answer, and an explanation is in order: Back in 1989, many scientists felt that Neptune's magnetic poles would be located near its rotational poles. As a result, astronomers commanded Voyager 2 to look for aurorae near the rotational poles, which turned out to be the wrong place. Later, they realized that the magnetic poles were far from the rotational poles. One image shows a brightening near one of the magnetic poles that may be an aurora. This area gives off ultraviolet light and is probably powered by the solar wind. We are not sure if it gives off visible light. Even if this area is Neptune's aurora, it is much weaker than Uranus' aurorae.

Middle Atmosphere

The middle atmosphere consists of the stratosphere, which starts at the tropopause near the 0.1 bar level, and extends to about the 1 µbar level. The temperature in the stratosphere rises from ~52 K at the bottom up to ~160 K near the top. A cross-section view of the middle and lower atmosphere is shown in Figure 2.2.

What is the composition of the stratosphere? As it turns out, over a dozen different gases are present. See Table 2.2. One compound of special interest is hydrogen cyanide (HCN) which is present in small quantities. HCN does not come from Neptune's interior because the low temperature would cause it to condense before it reached the stratosphere. Astronomers believe that it originates in Neptune's stratosphere. One possible way that it forms is through the reactions:

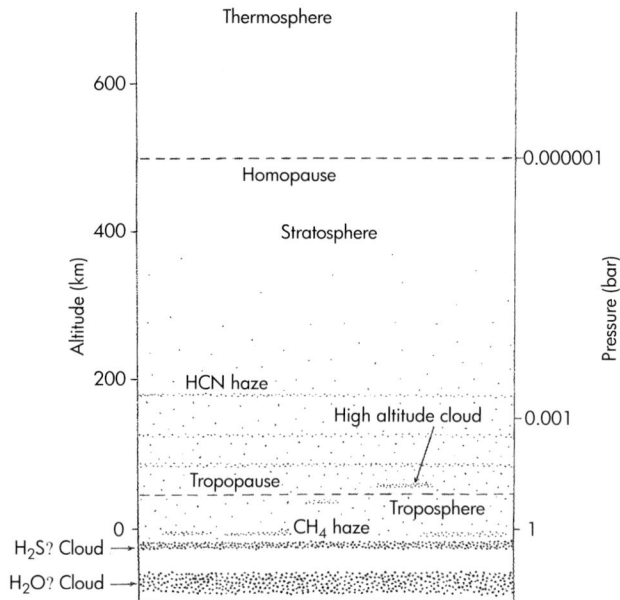

Figure 2.2. A diagram of Neptune's middle and lower atmosphere. The altitudes are with respect to the 1.0 bar level. Different levels of pressure are shown at right. (Credit: Richard W. Schmude, Jr.)

Table 2.2. Composition of Neptune's stratosphere

Component	Percentage by Volume
Hydrogen (H_2)	81 to 85
Helium (He)	15 to 19
Nitrogen (N_2)	0.3?
Methane (CH_4)	0.03
Acetylene (C_2H_2)	Trace
Ethene (C_2H_4)	Trace
Ethane (C_2H_6)	Trace
Methyl radical (CH_3)	Trace
Hydrogen cyanide (HCN)	Trace
Propyne (CH_3C_2H)	Trace
C_4H_2	Trace
C_3H_4	Trace
CH_3D	Trace[a]
HD	Trace[a]
$^{13}CH_3CH_3$	Trace[b]
Carbon Monoxide (CO)	Trace
Carbon Dioxide (CO_2)	Trace
Water (H_2O)	Trace
H_3^+	Trace
Methyl Radical (CH_3)	Trace

[a] We know that the D to H ratio $= 8 \times 10^{-5}$ to 1 from the amount of CH_3D and HD.
[b] We know that the ^{13}C to ^{12}C ratio $= 0.013$ to 1 from the amount of $^{13}CH_3CH_3$.

$$CH_3 + N \rightarrow H_2CN + H \qquad (2.7)$$

$$H + H_2CN \rightarrow HCN + H_2 \qquad (2.8)$$

Where did the atomic nitrogen (N) in the stratosphere come from? Two possible sources of N are: Triton and dissociation of atmospheric N_2 by galactic cosmic rays. Triton has a thin atmosphere, which is probably composed largely of nitrogen. Much of this nitrogen may escape from that moon due to its low gravity. If this occurs then some of it may end up in Neptune's atmosphere.

Galactic cosmic rays are fast moving protons and electrons which emanate outside of our Solar System. These particles are believed to react with molecular nitrogen (N_2) to produce atomic nitrogen (N). Voyager 2 detected N^+ ions when it made its closest approach to Neptune, and many of these may have made their way into the stratosphere. We are not sure of the amount of N_2 in the stratosphere. It may be as high as 0.6% or as low as trace amounts. As of 2006, astronomers have not detected stratospheric N and N_2; nevertheless, these species must be present in at least trace quantities.

There is some uncertainty in the make-up of Neptune's stratosphere. Its mean molecular weight is 2.38 grams/mole. This means that if we were to pick out one mole or 6.02×10^{23} atoms/molecules at random, their combined mass would be 2.38 grams. If we assume that the amount of N_2 is negligible, the percentages (by volume) of molecular hydrogen and helium are 81% and 19% respectively. If there is just 0.3% molecular nitrogen, the hydrogen and helium percentages change to 85% and 15% respectively. As a result, we need to know the amount of nitrogen in the stratosphere in order to better understand its chemical composition and thermal properties.

Figure 2.3 shows how the wind speed changes with latitude on Neptune. Many of these winds correspond to features in the lower stratosphere. Features near the equator rotate once every 19 hours, which is 400 m/s slower than the rotation period of Neptune's interior. Clouds at 30°S and 60°S have respective rotation rates of 17.6 and 15.4 hours and respective wind speeds of –230 m/s and 120 m/s. Like Uranus, Neptune's equatorial region has retrograde winds. The wind speeds in Figure 2.3 correspond to the ~0.1 bar level. The winds are slower at higher altitudes. In one study, astronomers estimated the wind speeds at the 0.38 mbar level to be only 60% of what they are at the 0.1 bar level. We know that the winds on Jupiter also change with altitude.

Small amounts of methane (CH_4) are in the stratosphere. The consensus is that methane makes up ~0.03% of the stratosphere by volume. This gas absorbs ultraviolet light from the Sun and several photochemical reactions occur. Several of these are listed in Chapter 1. Methane reactions are believed to lead to the formation of other hydrocarbons in the stratosphere. Photochemical reactions of methane also control the rate of haze production.

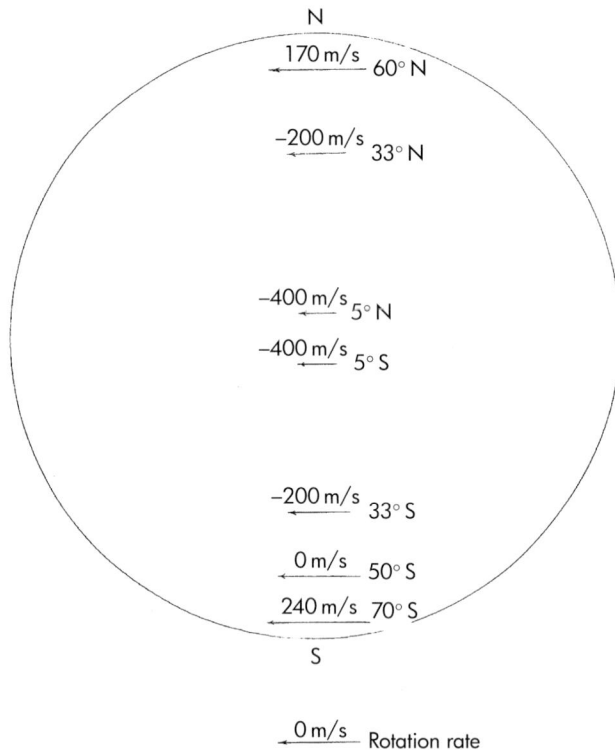

N

170 m/s   60° N

–200 m/s   33° N

–400 m/s 5° N

–400 m/s 5° S

–200 m/s   33° S

0 m/s   50° S

240 m/s   70° S

S

0 m/s   Rotation rate

Figure 2.3. Different latitudes of Neptune rotate at different speeds. The large circle is Neptune and the length of the arrow is proportional to the rotation speed. If the wind speed is positive, the wind is prograde; otherwise it is retrograde. Features moving with prograde winds rotate faster than Neptune's rotation rate of 16.108 hours; otherwise they rotate slower. (Credit: Richard W. Schmude, Jr.)

What is the nature of Neptune's haze? We have learned a great deal about its stratospheric haze from Voyager 2 data. In 1989, haze was present near the 12, 4 and 0.5 mbar levels. One group believes that the temperatures at these three levels are consistent with condensed ethane haze being present at the 12 mbar level, condensed hydrogen cyanide haze being present at the 4 mbar level and condensed C_4H_2 haze being present at the 0.5 mbar level. Neptune's haze in 1989 was thicker than Uranus' haze layer in 1986 and this may have been one reason why Neptune had more albedo irregularities in near-infrared light than Uranus during the 1980 s. During 1989, Neptune's haze layer absorbed a few percent of the visible light falling on it. We are not sure if the haze layer changes with Neptune's seasons. One group estimates an average haze production rate of 6 x 10^{-15} g/cm^2 per second based on 1989 data. The haze particles have diameters of ~0.4 μm and, as a result, it takes several years for them to fall through Neptune's stratosphere.

Like Uranus, methane is recycled on Neptune. Ultraviolet light reacts with methane in the stratosphere to produce large hydrocarbons which condense, producing haze particles. These particles trickle down to the troposphere. Once they heat up, they break apart into hydrocarbon gas. Some of this gas, in the form of methane, re-enters the stratosphere. The carbon cycle on Neptune is similar to that on Uranus.

When one looks out of a building and sees a tree, he or she looks through the air to the tree. In the same way, when we look at the bluish color of Neptune, we peer through the stratosphere and into the troposphere. See Figure 2.4. This is because the gases in the stratosphere are transparent just like the air in our atmosphere.

Neptune's stratospheric haze absorbs light and this plays an important role in the thermal structure in the stratosphere and may be a driving force for the formation and disappearance of stratospheric hazes.

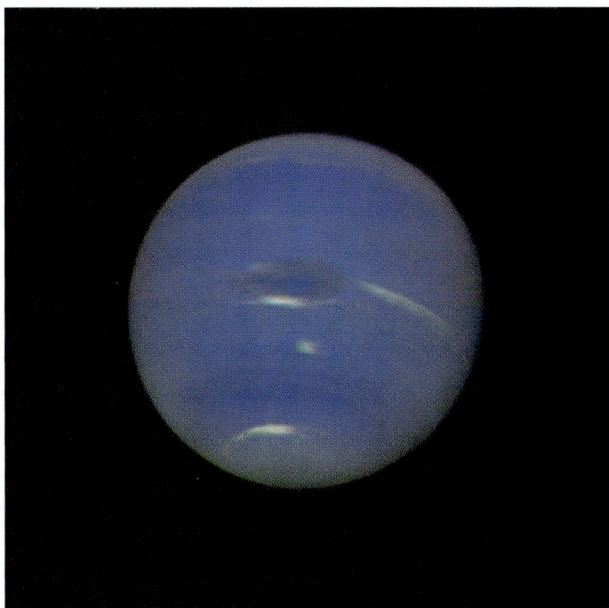

Figure 2.4. A Voyager 2 color image of Neptune taken at a distance of 16 million km. The Great Dark Spot (GDS) is in the center of the image. (Credit: Courtesy NASA/JPL-Caltech.)

The lower limit of the stratosphere is the tropopause, which is at the 0.1 bar level. The temperature of the tropopause varies with latitude and it may change with both the season and Neptune's rotation. The troposphere temperature at 42°S is 50 K. As on Uranus, the tropopause temperature does not change with altitude and is the coldest temperature in the stratosphere and troposphere. Compounds like water, ammonia and hydrogen cyanide condense before they reach the tropopause and are not able to cross from the troposphere to the stratosphere or vice-versa. See Figure 2.5.

Lower Atmosphere

The troposphere lies below the tropopause. Like Uranus, the temperature in Neptune's troposphere rises with increasing depth. When we look at Neptune through a telescope, we see the troposphere and its clouds. It contains ~3% methane, which is similar to Uranus. Other gases like water vapor and hydrogen

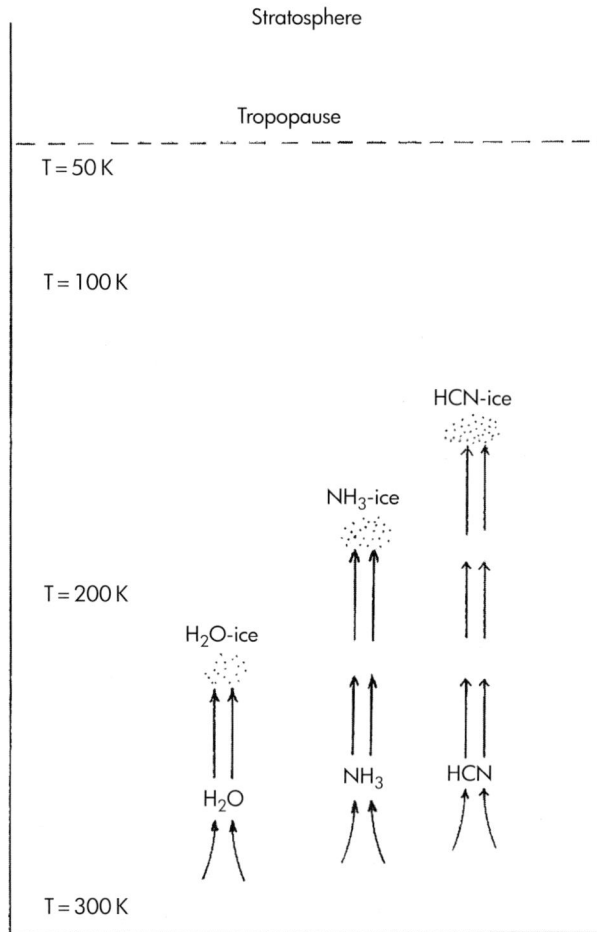

Figure 2.5. As H_2O, NH_3 and HCN rise, they encounter cold layers of gas and, as a result, these compounds turn to ice crystals before reaching the stratosphere. (Credit: Richard W. Schmude, Jr.)

sulfide are probably present. Ammonia is scarce or absent in the upper troposphere, but may be more plentiful at greater depths.

The amount of ortho- and para-hydrogen in Neptune's troposphere is near the
equilibrium value. This is similar to that of Uranus. Like Uranus, most of the
hydrogen has had time to reach the equilibrium ortho-to-para ratio.

At least three clouds layers are present in the troposphere. A thin methane cloud lies
near the 1.5 bar level. It may be patchy. Some visible light penetrates this layer.
Seasonal temperature changes and convection currents probably affect the thickness
and altitude of this cloud. A second cloud layer lies near the four bar level. It is believed
to be made up of condensed hydrogen sulfide along with small amounts of condensed
ammonia and possibly ammonia hydrosulfide (NH_4SH). This cloud is opaque to
visible light and has a bluish hue because it absorbs more red than blue light. A
third cloud layer probably lies at the $\sim$50 bar level and may be composed of water or a
solution of water, ammonia and other compounds. In addition to these clouds, there
may be lots of haze in the troposphere. Figure 2.6 shows some high altitude clouds on
Neptune casting shadows.

Microwave data can give us information about the troposphere at the 4 to
100 bar level. Like Uranus, Neptune's polar regions are releasing more microwaves
(wavelength between 1.3 and 3.6 cm) than the equatorial regions. A cloud at the
3 to 50 bar level may be thicker near the equator and hence would block more of
the microwaves than that near the polar regions.

Clouds

Neptune has several types of clouds which include dark oval features, bright
wispy clouds near dark ovals, high altitude clouds and the South Polar Feature.

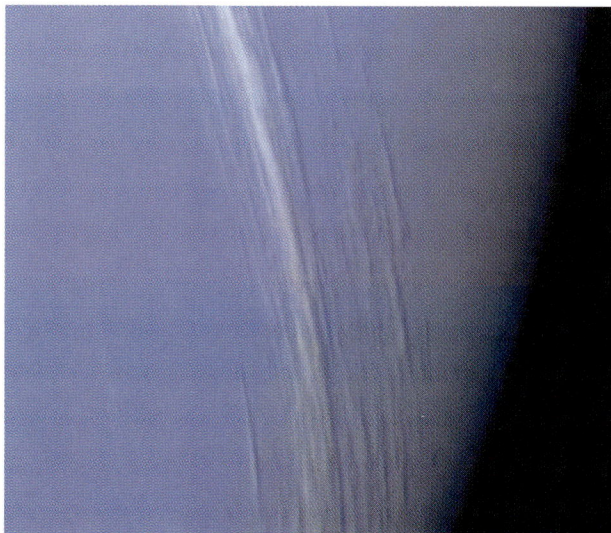

Figure 2.6. A Voyager 2 image of some high altitude clouds casting a shadow. These clouds are at
29°N and are about 50 km above the surrounding area. The sunlight comes from the lower left. The
left sides of these clouds are very bright because they are sloped towards the Sun. The shadows are
on the right sides of the clouds. (Credit: Courtesy NASA/JPL-Caltech.)

Table 2.3. Average dimensions of some of Neptune's well-known clouds

Cloud	Dimensions (km)		Area (10^6 km^2)
	East-West	North-south	
Great Dark Spot (GDS)	15,000	6,300	74
NGDS-32	11,000	4,300	36
NGDS-15	8,300	4,300	28
GDS companion[a]	10,000	2,300	18
NGDS-32 companion[b]	7,000	7,000	38
The Scooter[a]	3,000	2,000	4
South Polar Feature[c]	10,000	5,000	41
Dark Spot 2 (D2)	6,000	2,600	12
D2 bright cloud	2,000	2,000	3

[a] Dimensions are based on Voyager MeU filter.
[b] Dimensions are based on a Hubble Space Telescope image made in 0.89 μm light on Aug. 13, 1996.
[c] Dimensions are based on an Earth-based image made with the K filter (2.2 μm).

Astronomers have used photoelectric photometry to study the development of Neptune's clouds as related above; at least three cloud layers are in the troposphere. Some clouds lie in the lower stratosphere. Table 2.3 lists a few of these clouds and their sizes.

The large dark clouds include the Great Dark Spot (GDS) imaged in 1989 along with NGDS-32 and NGDS-15 (two more dark ovals) which were imaged from Earth in the 1990 s. These clouds are discussed first. Discussions of bright clouds will follow.

Figure 2.7 shows two Voyager 2 images of Neptune's GDS. This feature's contrast was higher in blue light than in green light. During 1989, ground-based observers were not able to image it in green light. The contrast for this feature was also low in Voyager 2 methane band images. The GDS underwent changes in location, shape and orientation during 1989. Two of these changes (shape and orientation) changed in a periodic manner. No other cloud has undergone such changes. Neptune's GDS is unique!

The latitude of the GDS changed in 1989. It was centered at 27°S on January 23, 1989, but it drifted northward to 17°S by late August, 1989. The rotation rate of the GDS around Neptune's axis changed from 17.95 hours when it was at 27°S to 18.4 hours when it was at 17°S. During 1989, the rotation rate changed gradually. The reason why the rotation period changed is because different latitudes rotate at different rates; or, in other words, the wind speed changes with latitude. See Figure 2.3. One small Jupiter spot (named S1 and later S2, when its rotation period changed) moved from 17.3°S to 16.1°S in early 1992, and this movement is probably what caused the change in rotation rate. This movement may be similar to what Neptune's GDS experienced.

The GDS underwent cyclic changes in its size and area. On one day it would have respective east-west and north-south dimensions of 12,000 km and 7,400 km but, four days later, the respective dimensions would be 18,000 km and 5,200 km. The shape would cycle every 7.9 days. These changes are shown in Figure 2.8. GDS' area also changed by 10% to 15% and the change was cyclic. The area changes were

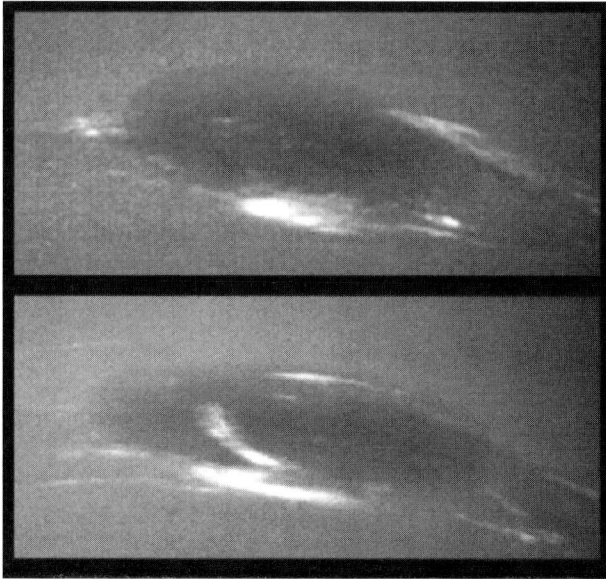

Figure 2.7. Two Voyager 2 images of Neptune's Great Dark Spot (GDS). The GDS goes through a cyclic change in shape every 7.9 days. The bright clouds lie above the GDS and undergo rapid changes in shape. (Credit: Courtesy NASA/JPL-Caltech.)

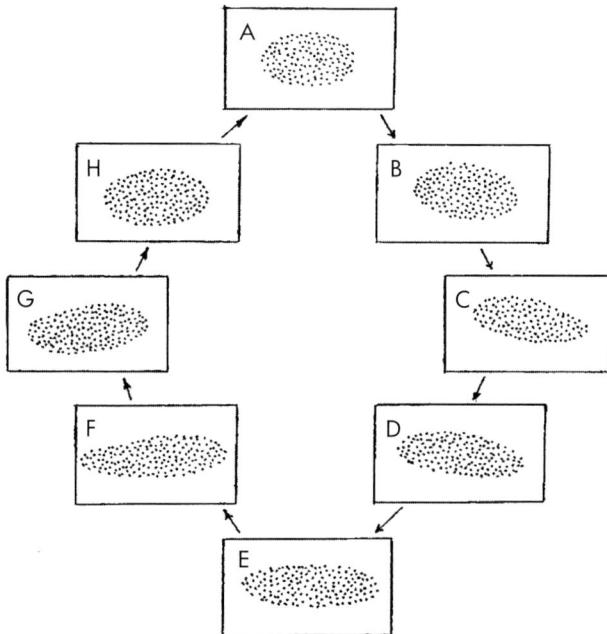

Figure 2.8. The dark oval represents the shape of the Great Dark Spot (GDS). The GDS changes shape over a 7.9 day period. Each box shows how the shape changes every ~24 hours. The length and width undergoes a periodic change; furthermore, the major axis swings a little. In frames A and E, the major axis of the GDS is parallel to the equator and the bottom of the page, but this is not the case in frames C and G. (Credit: Richard W. Schmude, Jr.)

small because when this spot got longer in the east-west direction it would get shorter in the north-south direction. No other feature on any other planet undergoes such an obvious cyclic change in shape as the GDS; however, oval BA on Jupiter becomes flatter as it passes the Great Red Spot. Oval BA probably changes because of the Great Red Spot. There are no known features causing the GDS to change in size and shape.

A third cyclic change that the GDS underwent in 1989 was its orientation. When the GDS had an east-west length of 15,000 km, its major axis was parallel to the equator, but this gradually changed and, after two days, its major axis was oriented $10°$ away from a line parallel to the equator. The orientation of the major axis continued to shift until after two more days it was once again parallel to the equator. The major axis, therefore, underwent a cyclic change in orientation. This cycle repeated itself every 7.9 days. See Figure 2.8.

What is the GDS? This is a difficult question to answer because Voyager 2 was only able to get a limited number of close-up images of this feature. The GDS is probably a vortex similar to Jupiter's Great Red Spot. Its winds swirl in a counter-clockwise direction. Voyager 2 did not yield any data on the wind speed inside of the GDS.

A second large dark spot (NGDS-32) developed in 1994 and lasted until at least August 1996. This feature was centered at $32.3°N$. Unlike the GDS, NGDS-32 remained near the same latitude. It did not undergo the large changes in shape that the GDS underwent. NGDS-32 shrunk from an area of 56 million square kilometers in November 1994 to 21 million square kilometers 1.8 years later. The GDS did not shrink or grow larger over time in 1989; however, it may have done so later. The rotation rate of NGDS-32 was near 17.27 hours, and it changed just slightly from late 1994 to late 1995. This change was probably caused by a slight change in latitude. Like the GDS, this feature had a bright companion cloud, but on its northern (or poleward) side. This companion is discussed later.

A third large dark spot, NGDS-15, developed in late 1995 or early 1996 and disappeared about two years later. Its average latitude was $15.2 \pm 1.4°N$ in 1996, but was $13.5 \pm 1°N$ in 1997. Its rotation period changed, and this led one group of astronomers to conclude that either this spot changed in latitude or that NGDS-15 was more than one spot that disappeared and reappeared. It had a lower contrast in blue light than NGDS-32, which may be due to its being at a lower altitude. Unlike the other two dark spots, NGDS-15 did not have an obvious bright cloud near it.

Two of the dark spots (GDS and NGDS-32) had companion clouds develop near their poleward edges. These clouds reflected lots of near-infrared light. As a result, they caused Neptune to brighten by up to 0.4 magnitudes in near-infrared light as they rotated into view. Each of the bright companion clouds is discussed here.

Throughout 1989, a bright cloud (the GDS companion) was centered near the southern (or poleward) edge of the GDS. This feature was made up of many small clouds with dimensions of $\sim$2,000 km (east-west) by $\sim$150 km (north-south). These clouds dissipated quickly, but new ones appeared. As a result, the shape of the GDS companion changed, but this cloud remained visible. It had the same rotation rate as the GDS even though it was centered at a latitude that was $7°$ farther south than the GDS. According to Neptune's wind profile, the GDS should have had a rotation period that was $\sim$15 minutes longer than its bright companion. One explanation for the rotation rate of the GDS companion is that it was caused by gases which were nearly saturated with methane and forced to higher and colder

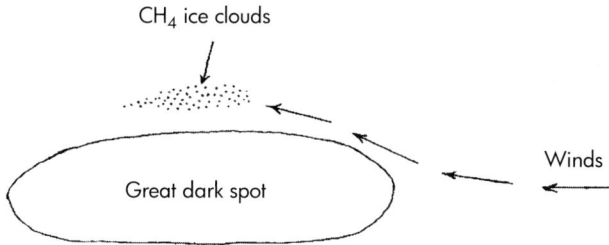

CH$_4$ ice clouds

Great dark spot

Winds

Figure 2.9. This figure shows a side view of the Great Dark Spot (GDS) illustrating how the white companion cloud forms. Winds force air containing lots of methane (CH$_4$) above the GDS where the temperature is lower. The lower temperatures cause some of the methane to freeze, forming the companion cloud. Since this cloud forms as a result of the GDS, it has the same rotation rate as the GDS. (Credit: Richard W. Schmude, Jr.)

altitudes near the GDS. The methane condensed into bright clouds at the high altitudes creating the bright companion feature. See Figure 2.9. If this is the case then, the GDS companion cloud was dependant on the GDS and would not exist without it.

The GDS companion resembles cirrus clouds on Earth. One group of astronomers believes that these clouds form near the tropopause (at the 0.1 bar level), when air is forced over the top of the GDS.

In many ways, the NGDS-32 companion is similar to its counterpart near the GDS. This feature was also on the poleward boundary of a large dark cloud. It had the same rotation rate as NGDS-32 even though its center was ∼6.4° farther north of the darker cloud. The NGDS-32 companion was not always bright in blue light, but was always bright in near-infrared light. It caused Neptune to brighten by ∼0.4 magnitudes in near-infrared light (wavelength = 0.89 μm) as it rotated into view, but it had little impact on Neptune's visible light brightness.

Small bright clouds were also on the eastern, western and northern edges of NGDS-32, but they came and left quickly. These clouds had areas of ∼300,000 square kilometers, which is comparable to that of the British Isles. Due to the limited resolution of Earth-based images, we are not sure if similar clouds formed around the other dark spots discussed above.

A bright cloud called "The Scooter" was distinct in many Voyager images. This cloud was centered at 40°S. Like the GDS companion, The Scooter appears as several long and thin clouds. It lasted for at least 81 days. This feature oscillated at latitudes of between ∼39°S and ∼41°S with a period of 22.5 days. It shifted also in longitude, but the shift is not well understood due to its long period of ∼121 days. The Scooter is believed to be below the 1.9 bar level.

In 1989, a second Dark Spot (D2), 12° south of The Scooter, was present. The size and area of D2 is shown in Table 2.3. Both the latitude and longitude of D2 oscillated. The latitude oscillated between ∼50°S and ∼55°S with a period of just over 36 days. D2 oscillated almost 47° in longitude with respect to a constant drift rate. The net effect of the longitude oscillation is that this feature's drift rate changed in a periodic manner. This behavior is similar to Jupiter's oscillating spots in 1940, 1941 and 1987.

A bright cloud often developed near the center of D2. See Figure 2.10. The area of this cloud changed. Its average area was ∼2 million square kilometers. It was largest when D2 was at its most northerly latitude (50°S).

Figure 2.10. A close-up image of Dark Spot 2 (D2) made by Voyager 2 in 1989. The banding is a sign of strong winds whirling around the center. This image covers an area of 7,000 by 10,000 km. (Credit: Courtesy NASA/JPL-Caltech.)

Astronomers have imaged the South Polar Feature (SPF) cloud on several occasions since the late 1980 s. This cloud is always within the south polar region; hence, its name. This cloud appears at the 68° to 74°S region and can disappear quickly. Like the other bright clouds, the SPF is made up of several smaller clouds. These clouds form and dissipate on a time scale of minutes to hours. The entire SPF can undergo large changes in a short time span. For example, it was a very large and bright feature on June 25, 2001, in 2.2 μm light, but one day later its area decreased by a factor of two and it became darker, and, after one more day, this feature was almost invisible. One group of astronomers reports that it was distinct in red light in June 2001. During 2000 and 2002, other astronomers imaged the SPF in several wavelengths of light. From this study, they reported that it reached the 0.27 to 0.17 bar level. In a 1989 image, shadows from the SPF are visible. Astronomers report that the SPF shadows indicate an altitude of 50 km above the thick cloud at the four bar level. This result is consistent with the SPF reaching the ~0.3 bar level.

Starting in the 1990 s, astronomers have imaged Neptune's bright cloud belts using near-infrared light. These areas appear bright in such light; hence, they are probably composed of high altitude hazes. They are not distinct in visible light, but, under the right conditions, one may be able to image them. I will limit my discussion to observations made in near-infrared light. The most distinct cloud belt is between 20°S and 50°S. A second belt lies near 40°N. The belt between 20°S and 50°S was in the form of two strips in 2000, but these merged into one thick strip by 2003. The belt near 40°N also widened during this time. One group of astronomers reports that the feature at 40°N reached the 0.023 to 0.064 bar level, whereas the corresponding feature between 20°S and 50°S reached the 0.10 to 0.14 bar level.

As described above, Neptune possesses a variety of clouds. These clouds can affect its appearance and brightness and play a role in its thermal properties. Like Uranus, Neptune's cloud structure may affect the amount of radio and thermal emissions given off. In one study, the south polar region gave off large amounts of infrared radiation with wavelengths of between 8 and 22 µm. Light at these wavelengths is sensitive to temperatures near 150 K. Apparently the south polar region is warmer than other areas. This warmth is due to the release of internal heat since that region receives less sunlight than the temperate and equatorial regions. The south polar region emits more radio waves with wavelengths of 1.3, 2.0 and 3.6 cm than do other areas. This is similar to the situation on Uranus.

Brightness Changes

Do the brightness and color of Neptune change? During the 1950s, astronomers asked themselves this same question. As a result, they began collecting brightness and color measurements of Neptune and have kept making these measurements up to the present time.

Neptune brightened by a large amount between 1965 and 2005. During this time, the sub-Earth latitude rose from ~0° to ~30°S; or, in other words, we saw more of Neptune's polar regions in 2005 than in 1965. Astronomers at Lowell Observatory reported brightness increases of 0.14 and 0.12 magnitudes in the Stromgren b and y filters between 1965 and 2005. (These filters are sensitive to light with wavelengths of 0.470 and 0.550 µm and are discussed in Chapter 5.) These results show that Neptune brightened and its color remained almost unchanged as the sub-Earth latitude moved south. These results are different from those of Uranus in at least two ways: (1) Neptune's color did not change much as it brightened, whereas Uranus became redder as it brightened, and (2) Neptune's y magnitude brightened by 0.12 magnitudes when the sub-Earth latitude went from 0° to 30°S, whereas Uranus brightened by only 0.03 magnitudes. These differences are due at least to different cloud structures and compositions of the two planets. The solar cycle may also affect Neptune's brightness.

Data collected by ALPO members between 1991 and 2006 are shown in Figure 2.11. These results were transformed to the Johnson V system. Neptune brightened by an average rate of 0.008 magnitude/year during this time interval. The ALPO data suggests that Neptune's B–V color index increased a little between 1993 and 2005; or, in other words, its color changed a little.

The seasonally averaged B–V values for Uranus (0.53) and Neptune (0.42) constitutes evidence that Neptune is bluer than Uranus. Neptune's stronger blue

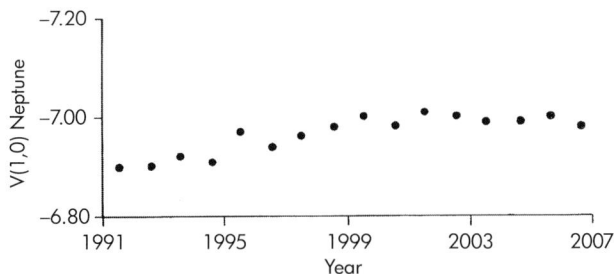

Figure 2.11. Normalized magnitude measurements of Neptune collected by members of the Association of Lunar & Planetary Observers (ALPO). The V(1,0) value is the magnitude that Neptune would have if it were 1.0 astronomical unit from both the Earth and Sun. (Credit: Richard W. Schmude, Jr.)

color may be due to the thick clouds near the four bar level resulting in a stronger blue color than similar clouds on Uranus.

Other Atmospheric Characteristics

Neptune's appearance changes in different wavelengths of light. Neptune often has bright patches and may have bright limbs in near-infrared light. I have just described the clouds on Neptune that reflect near-infrared light; in the next Section, I will describe Neptune's appearance in visible light.

In 2002, I had the chance to see Neptune through a 28 inch binocular telescope on top of Mount Evans in Colorado, which is at an altitude of 4,300 m (or 14,100 feet). Neptune had a bluish color with some limb darkening, but I was unable to see any other albedo features. The limb darkening may have been stronger on its northeast limb. A stronger limb darkening could be caused by one of several things, such as a change in haze abundance or a change in the thickness of the methane cloud layer near the 1.5 bar level. Those who have the proper electronic camera, filters and telescope are able to search for changes and irregularities in Neptune's limb darkening.

Why does Neptune have limb darkening? Figure 2.12 offers possible explanations for both limb darkening and limb brightening. Limb darkening occurs because more visible light is reflected by the disc center than the disc edge.

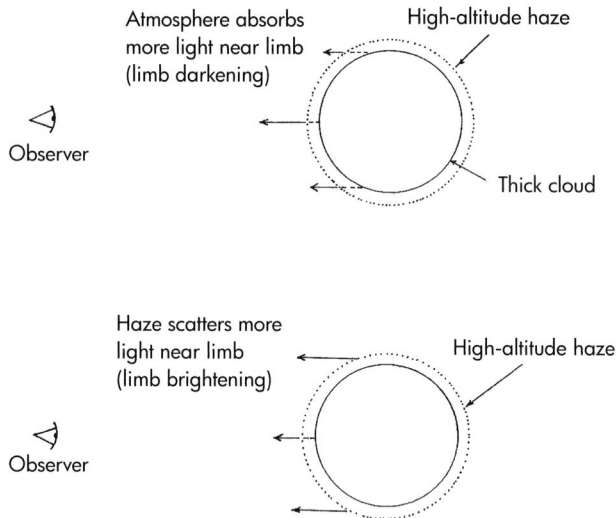

Figure 2.12. The way that haze scatters light will determine whether Neptune will have limb darkening or limb brightening. In the top frame, light is not scattered by the high altitude haze but is scattered by the thick cloud layer at lower altitudes. As a result, the light hitting the limb travels through a thicker portion of the atmosphere before being scattered. Since the atmosphere absorbs some of the light, this will cause limb darkening (short arrows). In the bottom frame, a different wavelength of light is used. In this case, the high altitude haze scatters this type of light. Since the light encounters a thicker part of the haze near the limb, more of the light is scattered, which causes limb brightening (long arrows). The length of the arrow is proportional to the amount of light reflected. (Credit: Richard W. Schmude, Jr.)

Much of the visible light striking the disc center is reflected by the thick cloud at the four bar level and most of it makes its way through the thin section of Neptune's atmosphere. Hence, the longer arrow in the top half of Figure 2.12 shows more light reflected by the disc center. The situation is different for light reflected near Neptune's edge. This is because the light must travel through a thicker portion of that planet's atmosphere before reaching the reflective cloud at the four bar level; furthermore, the reflected light must travel through a thicker gas layer before escaping Neptune. Essentially, more of this light is absorbed than light reflected by areas near the center, and this causes the limbs to appear darker; hence, limb darkening and the shorter arrow in the top half of Figure 2.12 symbolizing less reflected light.

Limb brightening occurs because the limbs reflect more light than the center as is shown in the bottom half of Figure 2.12. High altitude hazes may be the source of limb brightening. Essentially, Neptune's haze layers may be more efficient at scattering some wavelengths of light than others. Near the disc center these hazes are thin; hence, much of the light penetrates them and goes to lower depths where it is absorbed. As a result of this, very little light would be reflected and, hence, the small arrow in Figure 2.12 at the disc center. Due to the curvature of Neptune, the light near the limb strikes a thicker layer of haze and more of it is reflected back to Earth. Neptune had bright limbs (or limb brightening) in a Hubble Space Telescope image made in red light, but had dark limbs (or limb darkening) in an image made in blue light. This is because the haze scattered more red light than blue light. Therefore in some wavelengths of light, Neptune has limb darkening whereas in other wavelengths, it has limb brightening.

The most distinct feature on Neptune after limb darkening is a dark band at ~55°S. Two keen-eyed amateur astronomers observed this feature with the 1.0 meter Cassegrain telescope at PicduMidi observatory in the early 1990 s. These two also observed a bright spot near the south polar region, which may have been the SPF or another bright cloud. The dark band near 55°S also appears in a Hubble Space Telescope blue filter image.

The equatorial and polar radii of Neptune are listed in Table 2.1. Both radii correspond to the 1.0 bar level. These radii are consistent with an ellipticity of 0.017 ± 0.001 for Neptune. Two groups of astronomers report ellipticity values of 0.021 and 0.019 for Neptune near the 1.0 µbar level. These values are higher than the ellipticity near the 1.0 bar level.

Two different isotopes ^{13}C and ^{2}H (or deuterium) are present in Neptune's atmosphere. The ^{13}C to ^{12}C ratio is near 0.013 to 1, which is close to what it is on Earth. The D/H ratio is around 8.0×10^{-5} to 1. This is similar to the value found on Uranus.

Interior

The interiors of Neptune and Uranus have two differences, namely, (1) average density and (2) the release of internal heat. Neptune's density is almost 30% higher than that of Uranus. In fact, Neptune's average density is higher than some of Uranus' large moons. A second difference is that Neptune releases at least 10 times the amount of internal heat than that of Uranus. Deep convection currents from the hot interior probably transport much of Neptune's heat. Uranus either lacks these currents or its interior is cooler. One group reports that if one assumes a solar composition of uranium, potassium and thorium for Neptune, only about 0.01 to 0.02 Joules/m^2 of heat energy will be produced each second. This would be much less than the observed value of 0.4 Joules/(m^2 s). Where is the extra heat coming from? One study suggests that heat left over from Neptune's formation is being released into outer space and, as a result, its interior is slowly cooling.

Astronomers have used the same procedure for modeling Neptune's interior as what they used in modeling the interior of Uranus. Most of these studies suggest an interior similar to what is shown in Figure 2.13.

We believe that Neptune is made up of ~6% rock, ~84% ocean and ~10% hydrogen and helium. The hydrogen-helium layer extends from the 1.0 bar level down to a depth of ~5,000 km where the pressure reaches ~100,000 bar. Below the gas layer is the ocean layer which contains oxygen, nitrogen, carbon and hydrogen. As for Uranus, the term "ocean" does not imply a liquid layer; in fact, it may contain either unbonded oxygen, nitrogen, carbon and hydrogen or compounds like water, ammonia and methane. Below the ocean layer is a rocky core which may either be liquid or solid. There is a chance that substantial amounts of hydrogen are deep in Neptune's interior and, if this is the case, the ocean-gas boundary will be less than 5,000 km below the 1.0 bar level. If hydrogen exists in the deep interior, it may be in a metallic (instead of a molecular) state. Electrons can flow more freely in metallic hydrogen than in molecular hydrogen, and this may affect Neptune's magnetic field.

One must realize that the gas, ocean and rocky layers will have much higher densities than what they would have on Earth's surface. One group of astronomers predict average densities of 0.4, 4 and 10 g/cm^3 for Neptune's gas, ocean and rocky layers, respectively. The high densities are due to the extreme pressures inside of it.

One group of astronomers, who have computed models of the interiors of Uranus and Neptune, believe that Neptune's high density is due to its greater mass (compared to Uranus). Essentially material in Neptune is squeezed more than in Uranus, and this leads to Neptune being denser than Uranus.

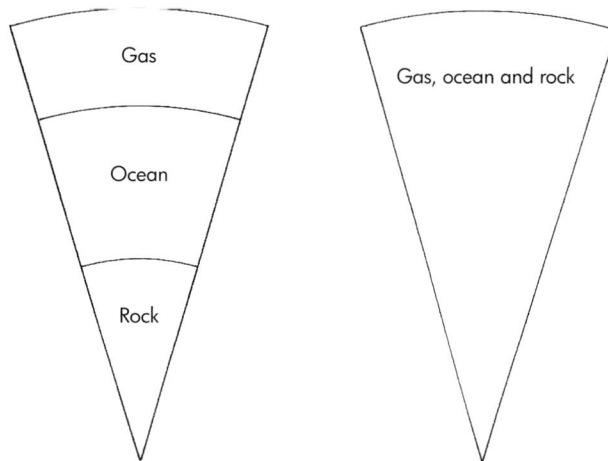

Figure 2.13. Two possible layouts of Neptune's interior are shown. The frame on the left shows distinct layers of gas, ocean and rock, while the frame on the right shows a homogeneous mixture of all three components. Neptune probably resembles the frame on the left. (Credit: Richard W. Schmude, Jr.)

Magnetic Environment

Voyager 2 data have revealed several characteristics of Neptune's magnetic field. Like Uranus, Neptune's magnetic field is not aligned with its rotation axis. See Figure 2.14. Neptune's magnetic dipole is tilted 47° to the rotational axis. As a result, the magnetic poles are located at tropical and temperate latitudes instead of polar ones; furthermore, they often point near the Sun. Neptune's magnetic field is weaker than Uranus' field. At the 1.0 bar level, the local field strength ranges from 0.1 to 1 Gauss. The field is stronger in the southern hemisphere. This is because the field is centered in the southern hemisphere and is about half-way between the center and the surface. As a result, the southern hemisphere is closer to the source of the field.

Like Earth and Uranus, Neptune's magnetic field is probably caused by a dynamo. Neptune's dynamo is believed to be close to the surface. Currents in the ocean layer may be the source of the dynamo. When one gets within about two planetary radii of the magnetic dynamo, higher order contributions (quadrupole and octupole) become important. In fact, higher order terms are larger in proportion to the dipole term for Neptune than on any of the other giant planets. If higher order terms were more important on Earth, a compass needle would often point in other directions besides the magnetic pole.

Since the higher order terms of Neptune's magnetic field become significant near the surface, one must resort to complex models to predict the location of aurora. One group of astronomers used a model of the magnetic field with quadrupole and octupole terms to predict locations of 35°S, 260°W and 60°N, 45°W for the aurora zones. As it turns out, if one assumes a dipole field, the two magnetic poles are several degrees from the centers of the predicted aurora zones.

Neptune possesses a magnetosphere that is similar in size to that of Uranus. Figure 2.15 shows the basic layout of the magnetosphere. The bow shock and magnetopause are 38.8 and 26.4 planetary radii respectively from Neptune's center. The magnetic pole was pointed towards the Sun and the incoming solar wind when Voyager 2 arrived near Neptune in 1989. The region near the pole is called the cusp region and Voyager 2 traveled through it as it approached Neptune.

Moons and rings can affect the distribution of electrons inside Neptune's magnetosphere. Figure 2.16 shows the distribution of high energy electrons inside of the magnetosphere. The concentration is highest near L = 7 Neptune radii.

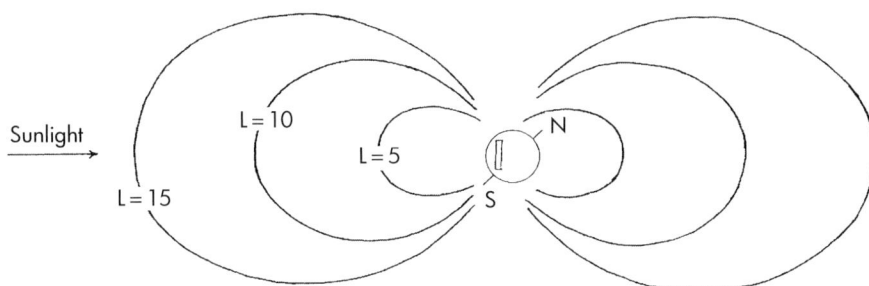

Figure 2.14. Orientation of Neptune's magnetic field and its rotational axis. Side views of the L = 5, 10 and 15 shells are also drawn. The N and S points define the rotational axis and the bar represents the magnetic field. (Credit: Richard W. Schmude, Jr.)

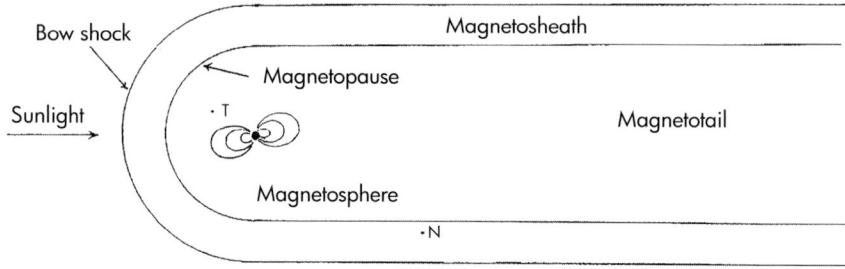

Figure 2.15. A layout of Neptune's magnetosphere. The points T and N are the positions of Triton and Nereid (two of Neptune's moons) at their closest point to Neptune. Neptune is the small ball. The L = 5, 10 and 15 shells of Neptune are shown also. (Credit: Richard W. Schmude, Jr.)

Throughout this Chapter, the units of L will be Neptune radii (24,764 km). Neptune's rings, moons and atmosphere absorb magnetospheric particles. One of Neptune's moons, Proteus, reaches a minimum L value of 5 and is responsible for the sharp drop at this area. The moon Triton is believed to play a role in the electron distribution as well but in a more complex way. Since that moon has an atmosphere, a gas torus may surround Neptune similar to Io's torus around Jupiter. The torus may interact with magnetosphere particles; however, more data are needed before firm conclusions can be made. At least two of the rings, Galle and Adams, absorb electrons as well. Absorption of electrons by the atmosphere needs more explanation. Figure 2.17 shows the L = 2 line. On one side of Neptune, it extends almost 40,000 km above the 1.0 bar level. At this distance, charged particles can move with little interference from the exosphere. On the other side of Neptune, the L = 2 shell does not extend as high and, as a result, Neptune's exosphere (and possibly its ionosphere) interferes with the charged particles thereby causing many of them to be absorbed. This is one reason why the high energy electron density drops near Neptune. There are also several small moons that have minimum L values ranging from 2 to 3 that may also be responsible for the low electron densities near Neptune.

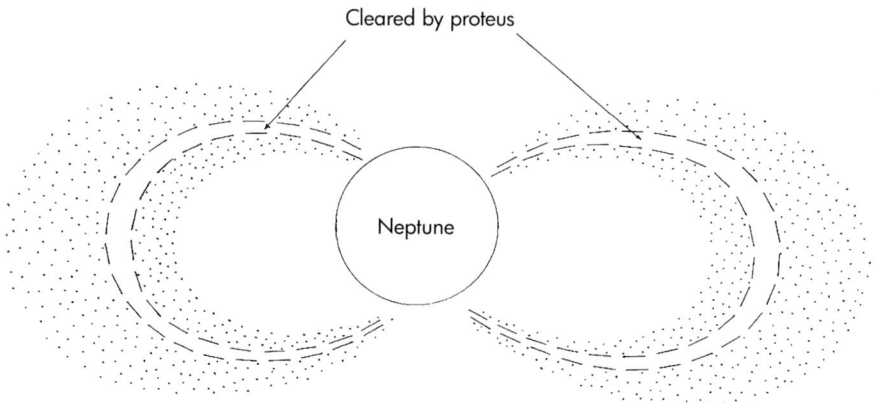

Figure 2.16. Distribution of electrons inside of Neptune's magnetosphere. Proteus, a moon of Neptune, has cleared out many of the electrons in the dashed area in the figure. Dots represent areas of high electron density. (Credit: Richard W. Schmude, Jr.)

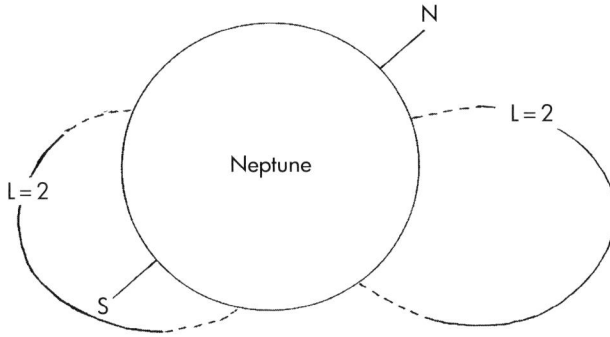

Figure 2.17. The L = 2 shell is really a three dimensional figure which extends around Neptune. The dashed portion of the curve contains part of the L = 2 shell and is within Neptune's upper atmosphere. As a result, many of the charged particles in the L = 2 shell will collide with the upper atmosphere as they drift in longitude and will be lost. (Credit: Richard W. Schmude, Jr.)

The magnetic field inside of Neptune's magnetosphere changes a little because of the trapped charged particles. Essentially, when electrons and ions move, they generate their own magnetic fields. Voyager 2 found that magnetospheric electrons altered the local magnetic field by just a few percent. The electrons in Earth's magnetosphere change the local magnetic field by a much larger amount.

What is the composition of the ions in Neptune's magnetosphere? This is difficult to answer because it depends on the energy of the ions and probably their location. Ion energies are equivalent to kinetic energies. Fast moving ions with the same mass have higher energies than slower moving ones. Almost all of the higher energy ions are H^+, with trace amounts of H_2^+ and He^+. There are practically no other high energy ions. The situation is different for ions with moderate energies. In addition to H^+, a sizable amount of heavier ions are present. These are probably N^+, which may come from Triton or a gas torus caused by Triton. The densities of moderate energy H^+ and N^+ are $\sim$0.4 ion/cm^3 and $\sim$1.6 ion/cm^3 at L = 1.7. The corresponding electron density is $\sim$2 electrons/cm^3. These densities are low compared to Jupiter and Saturn. Neptune's magnetosphere is considered to be almost empty compared to those of Jupiter and Saturn. Figure 2.18 shows how the density of N^+ changes with distance from Neptune. Unlike the case for high energy electrons, the N^+ density increases as one approaches Neptune.

There is some uncertainty of whether Voyager 2 entered the upper atmosphere of Neptune. Part of the problem here is that it is uncertain where the atmosphere

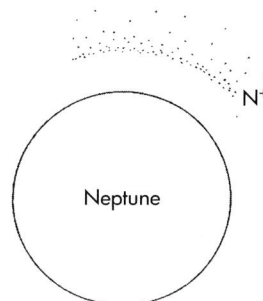

Figure 2.18. Voyager 2 passed close to Neptune in just one area and found a high density of N^+. Therefore, the N^+, shown as dots, is drawn only in one location near Neptune. The N^+ density increases with increasing magnetic field strength. It was denser at low L-shell values than higher L-shell values. We do not know the N^+ density within $\sim$5000 km of Neptune's cloud tops; hence, nothing is dawn there. (Credit: Richard W. Schmude, Jr.)

ends and the magnetosphere begins. Although some ions with moderate energies are at L = 1.7, lower energy electrons and ions are probably there as well.

There are three sources of ions in the magnetosphere: Solar wind, Triton and Neptune's atmosphere. Since there is 5% He^+ and almost no N^+ in the ion component of the solar wind, it is probably not the main source of high and moderate energy ions in Neptune's magnetosphere. Much of the N^+ is believed to be coming from Triton; after all, that moon has a thin nitrogen atmosphere and some of this gas is escaping into outer space. Finally, the presence of hydrogen cyanide (HCN) suggests that nitrogen is present in Neptune's stratosphere. The high energy ions in the magnetosphere probably come from Neptune's upper atmosphere since that area is believed to be made up almost entirely of atomic hydrogen. More data, however, are needed to pinpoint the source of ions in the magnetosphere.

Neptune has a magnetotail; however, we know very little about it. This is due to the unfavorable trajectory of Voyager 2 through it. Neptune's magnetic field rotates with the planet. At times, a magnetic pole is nearly pointed at the Sun, but, at other times, the magnetic poles are pointed far from the Sun. This changing orientation is believed to affect the shape of the plasma sheet and neutral boundary inside of the magnetotail. Figure 2.19 shows two situations. When the magnetic poles are facing away from the Sun, the situation is similar to what Uranus had in 1986; the plasma sheet in the magnetotail will probably have a curved shape. When a magnetic pole is pointed near the Sun, the plasma sheet takes the shape of a hollow cylinder pointed away from the Sun. The area on the inside of the plasma sheet will have the opposite magnetic polarity from the area outside of the sheet.

In 1989, Neptune emitted several types of radio waves, many of which had the same frequencies as AM radio stations here on Earth. The intensity of radio waves with frequencies of between 20 and 870 kilohertz, changed every 16.108 hours

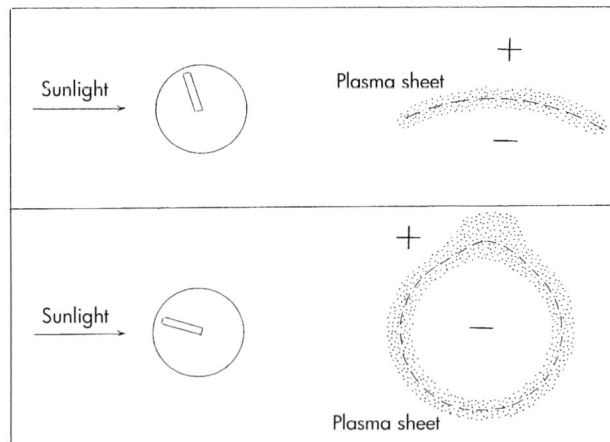

Figure 2.19. Cross-section views of Neptune's plasma sheet (dots) and neutral boundary (dashed lines). The orientation of Neptune's magnetic field to the Sun determines the shape of its plasma sheet and neutral boundary. If both of the magnetic poles point away from the Sun, the plasma sheet will have a curved shape, and a cross-section of it will look like the right portion of the top frame. If one of the magnetic poles points near the Sun, as is shown in the bottom frame, the plasma sheet will have the shape of a hollow cylinder, and a cross-section of it will look like the right portion of the bottom frame. (Credit: Richard W. Schmude, Jr.)

during the Voyager 2 encounter. This is believed to be the rotation rate of the planet's magnetic field and interior. Voyager 2 data reveal that Neptune emitted several bursts of radio waves, called whistlers. These bursts would start with a high frequency but, as time passed, the frequency dropped, hence the name whistler. Lightening can cause this type of radio burst. One group used the number of whistlers to estimate that ~100 large lightening events occur every hour, which is similar to the rate on Uranus.

Rings and Arcs

Neptune possesses five continuous rings; two of them are wide and the others are narrow. There is also one ring that is discontinuous. These rings and their characteristics are summarized in Table 2.4 and are illustrated in Figure 2.20. Figure 2.21 shows a Voyager 2 image of the brightest rings. All of Neptune's rings have low optical depths, which means that they do not block out much light when they occult a star. The more opaque something is, the higher its optical depth and the lower its transmission. Typical optical depths of Uranus' rings are ~0.3, whereas Neptune's rings have optical depths of ~0.003. One can think of Uranus' rings blocking out about as much light as lightly tinted glass, whereas Neptune's rings block about as much light as a well-coated and clean lens. Due to the low optical depth of Neptune's rings, they are difficult to study and, as a result, we know less about them than we do of Uranus' rings.

Two of Neptune's rings (Lassell and Galle) are wide, but since they reflect almost no light, they are not visible from Earth. There is a chance that when the sub-Earth latitude approaches 0°. These rings may cast a faint shadow on Neptune and may scatter enough light to show up in images. Neptune will reach its next equinox in about 2045.

One convenient way of describing Neptune's rings is in terms of the total amount of light removed from a light source during an occultation. There are two ways of doing this–equivalent width and equivalent depth. Equivalent width: For a narrow ring of radial width W, uniform transmission and tilted at an angle B with respect to the detector (or observer) one can write:

$$E = W \times (1 - f) \times \text{Sin}(B) \qquad (2.9)$$

Table 2.4. Characteristics of Neptune's rings

Ring	Average Distance from Neptune's center (km)	Period of revolution (days)	Width (km)	Composition[a]
Galle	42,000	0.239	~4,000	?
Le Verrier	53,200	0.341	~110	D, LP
Lassell	55,500	0.364	~4,000	?
Arago	57,500	0.383	~600	D, LP
Unnamed	62,000	0.429	~100	?
Adams	62,900	0.439	~30	D, LP

[a] D = dust; LP = large particles

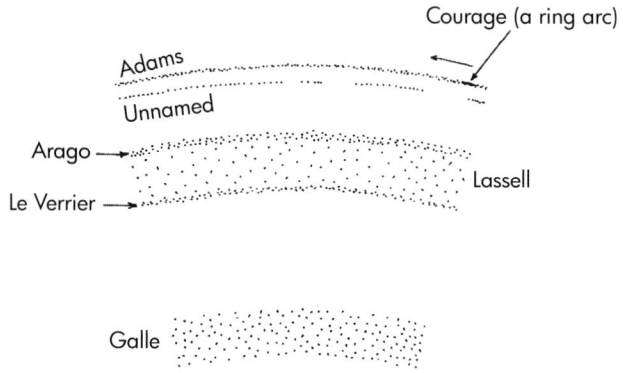

Figure 2.20. Positions and relative widths of Neptune's six rings. (Credit: Richard W. Schmude, Jr.)

Figure 2.21. Voyager 2 image of several of Neptune's rings. The ring arcs which are behind the planet are not visible in this image. (Credit: Courtesy NASA/JPL-Caltech.)

where E is the equivalent width, f is the line-of-sight transmission and Sin is the Sine function. The transmission is the fraction of light that an object transmits; it is nearly the opposite of optical depth. As f approaches 1.00, the optical depth approaches 0. Tinted glass typically has a transmission of ~0.50, whereas a coated and clean lens has a transmission of ~0.99.

Equivalent Depth: For well-resolved rings of known widths and optical depths, one can also determine the equivalent depth (A) which is:

$$A = W \times \tau \qquad (2.10)$$

where τ is the optical depth. As the optical depth and width increase, the value of A and E increase. Keep in mind that as f decreases, τ increases in equation 2.9. Therefore, the higher the values of E and A are for a ring, the more light that it will block during a stellar occultation. One must remember that E and A change with different wavelengths of light. Most occultation studies from Earth were carried out in near-infrared light, whereas Voyager 2 occultation measurements were carried out in ultraviolet light.

Table 2.5 lists the equivalent widths for several rings and arcs including two of Uranus's rings. As one can see, even Uranus' narrow 6 ring has an equivalent width that is much higher than either the Adams or Le Verrier rings. Neptune's ring arcs have a comparable equivalent depth to Uranus' 6 ring.

What are the characteristics of the ring particles? The large particles are dark and reflect only a few percent of the light falling on them. This is similar to the particles in Uranus' rings. There is some evidence that Neptune's large ring particles have a reddish color. These particles are probably composed of water ice along with dark material. The material may be dark as a result of the constant bombardment of high-energy charged particles on the rings.

Unlike most of Uranus' rings, Neptune's rings contain large amounts of dust. The high dust levels may be due to the larger distance of the rings from Neptune, combined with that planet's higher mass. The high mass forces Neptune's corona to lower altitudes and, as a result, the corona is thinner near the rings and drag forces are reduced. (The corona is the outer layer of Neptune's exosphere.) Alternatively, the high dust levels in the rings may be due to a large dust source that is not present in Uranus' rings.

Table 2.5. Equivalent widths and equivalent depths for some of Neptune's rings and ring arcs and similar data for selected rings of Uranus based on Voyager 2 measurements

Ring	Equivalent width (km)	Equivalent depth (km)
Le Verrier	<0.15[a]	0.7[b]
Liberté arc	1.6	0.7[c]
Adams ring	<0.15[a]	–
Galle ring arc	0.6[a]	–
Unnamed	<0.2[a]	–
Arago	<0.2[a]	–
Epsilon ring (Uranus)	40	60
6 ring (Uranus)	0.6	0.7

[a] For a wavelength of 2.2 μm.
[b] For a wavelength of 0.26 μm.
[c] Average of ultraviolet measurements at two different wavelengths.

Shepherd moons are probably what keep Neptune's rings in place. Since we do not know the exact widths of the rings, it is difficult to come up with mathematical models explaining their stability.

There is not enough information to determine an accurate mass of Neptune's rings. We have little information on the distribution of particle sizes. One group of astronomers suggests a mass of 0.0001 times that of Naiad's mass for the entire ring system. This comes out to be ~2 x 10^{13} kg.

In the following Section, I will give brief descriptions of each of the rings, starting with the innermost one.

Galle Ring

The Galle ring is wide with a very low particle density. As a result of this, it reflects almost no light, and it has a very low optical depth. An arc inside of this ring blocked out starlight in July 1986. This feature was 42,200 km from Neptune's center. It has an equivalent width of 0.6 km, which is close to that of Uranus' 6 ring. Astronomers have not detected this arc since 1989. The Galle ring is just a little brighter in forward-scattered light than in backscattered light. This means that it has a lower percentage of dust than some of the other rings around Neptune. This may be due to its closer distance to Neptune where drag forces from the corona are higher than for the more distant rings.

Le Verrier Ring

The Le Verrier Ring has a uniform brightness and is continuous at all longitudes. Its width is around 110 km. Since it is much brighter in forward-scattered light than in backscattered light, we know that it contains a large percentage of dust. In spite of this, astronomers used the Hubble Space Telescope and large Earth-based telescopes to image this ring in backscattered light. This ring is less than 30 km thick and lies close to Neptune's equator. Part of it has a 52-to-53 resonance with the moon Despina; therefore, this moon may play a role in constraining it. This ring was much brighter in a recent Hubble Space Telescope image than what it was in 1989. This brightening may be due partly to the opposition surge since the solar phase angle of Neptune is always ~2° or less from Earth, whereas it ranged from 8° to 160° in Voyager 2 images. The Le Verrier Ring is inclined to the plane containing the Adams and Galle Rings by 0.03°, and lies close to Neptune's equatorial plane.

Lassell Ring

Like the Galle Ring, the Lassell Ring is very wide and it absorbs less visible light than a clean piece of glass. It is continuous at all longitudes. This ring may have three thin arcs inside of it since a triple occultation was observed on April 18, 1984. At that time, these arcs had an equivalent width of ~1 km, which is similar to the arcs in the Adams ring. However, astronomers have not detected these features in later studies. There is some evidence that the Lassell Ring has a different particle size distribution than the Adams and Le Verrier Rings.

Arago and the Unnamed Ring

The Arago Ring is narrow and continuous. Voyager 2 imaged it at a high solar phase angle. This ring is very faint and we know very little about it.

A very faint unnamed ring lies at the same orbital distance as Neptune's moon Galatea. It is narrow and noncontinuous. Galatea undoubtedly plays some role in constraining it, and may even serve as a source of particles for it.

Adams Ring

The Adams Ring is the outermost ring and was the first one discovered. Particles in it may lie near the outer limit of Neptune's Roche limit. This ring is continuous at all longitudes, has a width of around 30 km and is less than 30 km thick. In 1989, several bright arcs were inside of it, and all of them were within a span of ~40°. At high solar phase angles, this is the second brightest ring behind the Le Verrier Ring. The Adams Ring contains a fair amount of dust. Astronomers used the Hubble Space Telescope to image this ring in 2005 and found that it was similar to what it was in 1989 except for its arcs.

Ring Arcs

The brightest portion of Neptune's ring system is the arcs which lie inside of the Adams Ring. They have a higher percentage of dust than the remainder of the Adams Ring. The optical depths of the arcs in near-infrared and ultraviolet light were ~0.1 in 1989. One can think of them as absorbing as much light as lightly tinted glass. The arcs discovered in 1989 are named Courage (French for courage), Liberté (French for liberty), Egalité 1 (French for equality), Egalité 2 and Fraternité (French for brotherhood). Courage is the leading arc followed by Liberté, Egalité 1, Egalité 2 and Fraternité. Characteristics of these arcs in 1989 are listed in Table 2.6. A diagram of the arcs is shown in Figure 2.22. Three of the arcs (Liberté, Egalité 1 and Egalité 2) have widths of ~15 km. We are not sure of the widths of the other two arcs, but they are probably close to 15 km. The Liberté arc probably has a faint shoulder on its exterior side, and the Egalité 1 arc may have a faint interior component with an optical depth of ~0.008. Figure 2.23 shows an image of three of the arcs.

In August 1989, the brightest arc in forward-scattered light was Egalité 2, but in backscattered light Liberté and Egalité 2 had nearly the same brightness. This suggests that Egalité 2 may have had a larger fraction of dust than Liberté. The situation may be different today. Courage and Liberté were much fainter in 2005

Table 2.6. Characteristics of Neptune's ring arcs in the Adams Ring in 1989

Arc	Length (degrees)	Length (km)	Width (km)	Composition
Courage	1	1000	15?	D, LP
Liberté	4.1	4500	~15	D, LP
Egalité 1	~1	~1000	~15	D, LP
Egalité 2	~3	~3000	~15	D, LP
Fraternité	9.6	11,000	15?	D, LP

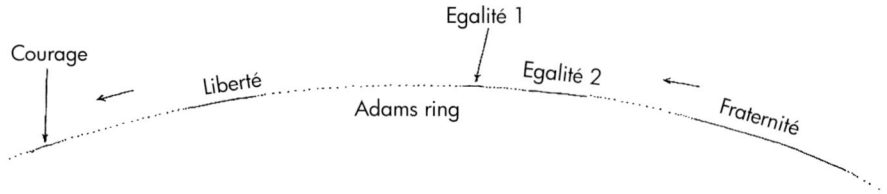

Figure 2.22. Diagram showing the positions and nearly true shapes of Neptune's ring arcs in 1989. Low density regions of the Adams ring lie between various ring arcs. (Credit: Richard W. Schmude, Jr.)

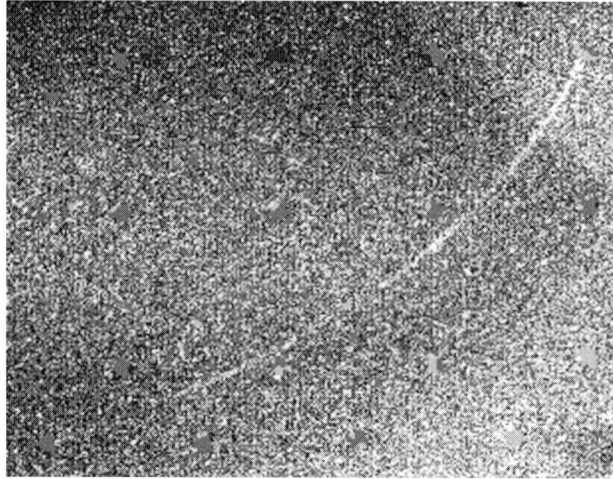

Figure 2.23. A Voyager 2 image of the ring arcs at a distance of 8.6 million km. The ring arcs are the brighter areas of the outer ring. Due to the limited resolution and great distance of Voyager 2 from the rings, the rings and arcs appear to be thicker than what they are in reality. (Credit: Courtesy NASA/JPL-Caltech.)

than in 1989; however, Egalité 2 and Fraternité have maintained a similar brightness to what they had 16 years earlier.

What forces are responsible for the arcs and their changing appearance? This is difficult to answer due to our lack of data. The arcs seem to resemble Saturn's F Ring. Parts of the F Ring have dimmed and brightened on short time scales like Courage and Liberté. One group of astronomers suggests that the many strands in Saturn's F Ring may be a single spiral arm which is constantly evolving. Several moons orbit near Saturn's F Ring, and at least one small moonlet with a diameter of less than one kilometer is believed to be within that ring. Neptune's arcs are probably constrained by the gravitational forces of one or more nearby moons. We know that Galatea has some influence because it has a 43-to-42 resonance with the arcs, meaning that it makes 43 trips around Neptune for every 42 trips that the arcs make and, hence, passes the arcs every ~18.4 days.

Dust

A faint ring of dust lies near Neptune's equatorial plane. Its maximum density is ~1 particle per $100 \, \mathrm{m}^3$. ($100 \, \mathrm{m}^3$ is about the volume of a school bus.) The dust density is higher than for Uranus. This may be due to Voyager passing closer to

Table 2.7. Names, orbital, and physical characteristics of Neptune's moons

Name	Distance [a] (km)	Orbital Period (days)	I (°) [b]	Radius (km)	Mass (10^{18} kg)	Density (g/cm^3)
Naiad [c]	48,200	0.294	4.74	48×30×26	0.2	1.5?
Thalassa [c]	50,100	0.311	0.21	54×50×26	0.4	1.5?
Despina [c]	52,500	0.335	0.06	90×74×64	3	1.5?
Galatea [c]	62,000	0.429	0.06	102×92×72	4	1.5?
Larissa [c]	73,500	0.555	0.20	108×102×84	6	1.5?
Proteus [c]	117,600	1.122	0.03	220×208×202	60	1.5?
Triton [d]	354,800	5.877	157	1,352.6	21,400	2.06
Nereid [e]	5,513,400	360.	7.23	170	30	1.5?
Halimede [f]	15,686,000	1,728	134	17	0.03	1.5?
Sao [f]	22,452,000	2,960	48	8	0.003	1.5?
Laomedeia [f]	22,580,000	2,984	35	9	0.004	1.5?
Neso [f]	46,570,000	8,841	132	11	0.01	1.5?
Psamathe [f]	46,738,000	8,889	137	9	0.005	1.5?

[a] The average distance or semi-major axis
[b] I = inclination [c] The distances, periods and inclinations for these moons are from Jacobson and Owen (2004), while the radii are from Karkoschka (2003). The mass values are computed from the assumed density value and a tri-axial ellipsoid geometry. [d] All characteristics for this moon are from Cruikshank et al (1995), p. 809. [e] The distance, period and inclination for this moon are from Kelly (2006), p. 26. The radius is from Cruikshank et al (1995), p. 687. The writer computed the mass from the assumed density and a spherical geometry.
[f] The distances, periods of revolution and inclinations are taken from Kelly (2006). I computed the radii of these moons from an assumed spherical geometry and an albedo that is equal to Nereid's albedo. I computed the mass value from the assumed density value and a spherical geometry.

Neptune. Neptune's dust probably comes from material ejected from the moons and ring particles. The equatorial dust belt may have a surface density as high as 10^{-8} g/cm^2 and may extend to ~170,000 km above the 1.0 bar level. This dust layer is at least 200 km thick, and is slightly inclined to Neptune's equatorial plane. This inclination may be caused by Triton. A very thin cloud of dust extends beyond this dust belt and may cover the entire Neptune system. Voyager 2 detected dust near Neptune's south polar region.

Some of this dust may be ionized, which means that the dust has a net electrical charge. Sunlight and impacts by fast moving sub-atomic particles in the magnetosphere can cause ionization. Ionized dust particles may even move with Neptune's magnetic field.

Satellites

Neptune has 13 known moons. Triton, the largest, is in a class by itself. The other 12 are either in the collision fragment category or the captured object category. Table 2.7 summarizes the orbital and physical characteristics of these moons, while Table 2.8 summarizes their photometric constants.

Table 2.8. Photometric constants of Neptune's moons

Name	V_o	V(1,0)	Geometric Albedo wavelength = 0.54 μm
Naiad[a]	23.91	9.20	0.072
Thalassa[a]	23.32	8.61	0.091
Despina[a]	22.00	7.29	0.090
Galatea[a]	21.85	7.14	0.079
Larissa[a]	21.49	6.78	0.091
Proteus[a]	19.75	5.05	0.096
Triton[b]	13.45	−1.26	0.72
Nereid[c]	19.0	4.4	0.26
Halimede[c]	24.0	9.4	0.26?
Sao[d]	25.8	11.1	0.26?
Laomedeia[d]	25.4	10.7	0.26?
Neso[d]	25.1	10.4	0.26?
Psamathe[d]	25.5	10.8	0.26?

[a] The V(1,0) and geometric albedo values are from Karkoshka (2003). I computed the Vo values form the V(1,0) values. [b] The V(1,0) value was computed from magnitudes reported by Hicks and Buratti (2004) and assuming a solar phase angle coefficient of 0.03 magnitudes per degree. I computed the Vo and geometric albedos from the V(1,0) values.
[c] The V(1,0) value is from Grav et al (2004). The geometric albedo is assumed to equal that of Nereid. [d] I computed the Vo value by adding 0.4 magnitudes to the Ro value reported in Holman et al (2004).

Triton

Triton is Neptune's largest moon and it is the seventh largest moon in the Solar System. As a result of Voyager 2, we know that Triton has an atmosphere and a variety of surface features. Before I discuss these characteristics, I would like to discuss its orbit and seasons.

Triton's Orbit and Seasons

Triton is not considered a "regular moon" because it has a retrograde orbit. It moves in the opposite direction that Neptune rotates and revolves. See Figure 2.24. This unnatural characteristic is probably due to its being captured by Neptune's gravitational pull. In order for such a capture to take place, Triton must have lost energy when it approached Neptune. One group of astronomers suggests that it was a binary object, much like Pluto and Charon, and when it approached

Figure 2.24. Triton rotates and revolves in a direction opposite to that of Neptune; hence, Triton has a retrograde orbit. (Credit: Richard W. Schmude, Jr.)

Table 2.9. Additional physical characteristics of Triton

Characteristic	Value
Ellipticity	0.0016 ± 0.001
Surface Temperature	38 K
Atmospheric pressure at surface	14 µbar (1989); 19 µbar (1997)
Surface area	2.30×10^7 km^2
Compounds identified in the atmosphere	N_2, CH_4
Compounds identified on the surface	Ices of N_2, CH_4, CO, CO_2, H_2O and C_2H_6
Orbital eccentricity	0.000016

Neptune, it was captured but its partner was ejected. Like the large moons of Uranus, Triton has a synchronous rotation, and, hence, the same side of it always faces Neptune. It also rotates in the opposite direction of Neptune. Additional physical characteristics of Triton are listed in Table 2.9.

Triton's axis is not parallel to Neptune's axis; hence, it experiences different seasons from those of Neptune. In Figure 2.25A, the sub-solar latitude of Neptune is at 30°S, which is the latitude where sunlight hits most directly. Since Triton's axis is inclined differently than Neptune's, its sub-solar latitude is close to 50°S in the Figure. At other times, however, Triton's axis can be pointed in a direction where the sub-solar latitude is at 10°S when this latitude is 30°S for Neptune. See Figure 2.25B. Therefore, it can experience either more extreme or less extreme seasons than Neptune. The situation, however, gets even more complex. Triton's orbit intersects Neptune's equatorial plane at two points called nodes. As it turns out, the line defined by the nodes makes a trip around Neptune every 688 years,

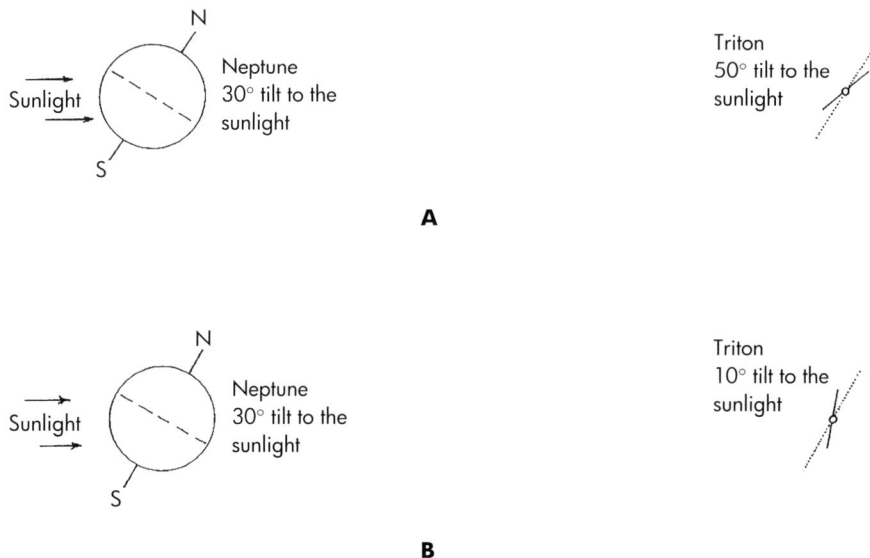

A

B

Figure 2.25. Triton's axis is not parallel to Neptune's N-S axis and, hence, it experiences different seasons. The dashed line near Triton is parallel to Neptune's axis. In the top frame (A), Triton's axis is tilted at a different angle and, as a result, the sub-solar latitude on it is 50°S, whereas it is 30°S for Neptune. In the bottom frame, Triton's axis is tilted in such a way that the sub-solar latitude is just 10°S, whereas it is 30°S for Neptune. (Credit: Richard W. Schmude, Jr.)

Table 2.10. Seasons on Triton: 1500 to 2100 AD[a]

Year	Season
1520	Hot southern summer
1570	Equinox
1610	Mild southern winter
1650	Equinox
1680	Very mild southern summer
1710	Equinox
1750	Very mild southern winter
1780	Equinox
1820	Mild southern summer
1860	Equinox
1910	Cold southern winter
1950	Equinox
2000	Very hot southern summer
2040	Equinox
2090	Cold southern winter

[a] Hot and cold are used as relative terms. A very hot summer may refer to a temperature of 40 K instead of 38 K. This table is based largely on Figure 11 in Cruikshank et al (1995), p. 1059.

or, in other words, Triton's axis precesses. As a result of axis precession and Neptune's revolution around the Sun, Triton's seasons can range from being very mild to extreme. Table 2.10 lists descriptions of a few of its seasons over the last few hundred years and for the years 2040 and 2090. As can be seen, that moon's southern hemisphere went from a cold southern winter in 1910 to a very hot southern summer 90 years later. (As explained in Table 2.10, "hot" and "cold" are relative terms.) This is different from the time period from 1610 to 1820 when the seasons were mild and less extreme than those on Neptune. At times, the sub-solar latitude can be as far south as 52°S and as far north as 52°N. During the year 2000, the sub-solar latitude was near 52°S. Hence, during the early 21st century, Triton's southern hemisphere experienced a very hot summer. The last time the sub-solar latitude reached 50°S was around 1340 AD. The length of Triton's seasons does not change much as a result of precession. Its seasons are each about 40 years long.

One group of astronomers estimates a seasonal temperature change of a few degrees Kelvin on Triton's surface. Such a change can cause the surface pressure to change by over a factor of two. Temperature changes due to Triton's rotation are less than 1 K. Small temperature changes due to cloud cover may occur over small parts of the surface. Dark areas will warm up more than brighter ones.

Triton's Atmosphere

Triton has a thin atmosphere, which is divided into four parts: troposphere, thermosphere, ionosphere and exosphere. Figure 2.26 shows the four parts.

The exobase is at an altitude of ~930 km. Atoms above the exobase are more likely to escape from Triton than those below the exobase. One group of astronomers reports that Triton loses about five million kilograms of hydrogen and about seven

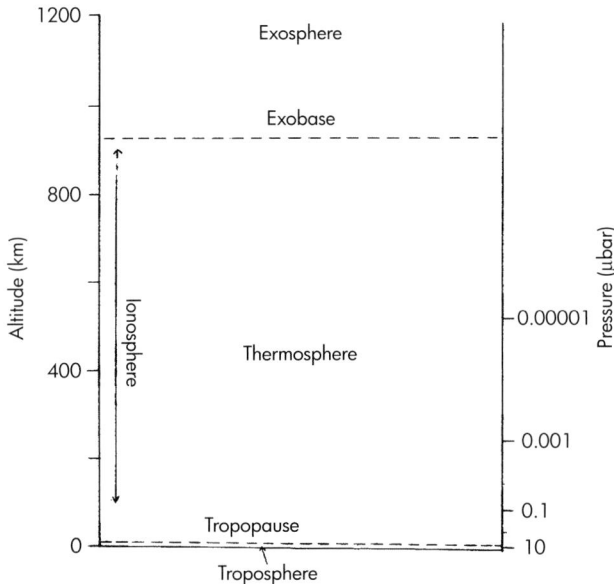

Figure 2.26. A diagram of Triton's atmosphere. The altitude above the surface is shown at the left and the atmospheric pressure at different altitudes is shown at the right. (Credit: Richard W. Schmude, Jr.)

million kilograms of nitrogen each year. The exosphere lies above the exobase and is probably made up of atomic nitrogen. The temperature of the exobase is near 100 K.

The thermosphere lies between the tropopause (altitude = 8 km) and the exobase. Its temperature rises from ~37 K at the tropopause to 100 K at an altitude of 500 km. The thermosphere gives off ultraviolet light, which is called airglow. The airglow is probably caused by ultraviolet light from the Sun and/or charged particles in Neptune's magnetosphere colliding with Triton's upper atmosphere. The intensity of the airglow probably depends on its location in the magnetosphere and the solar cycle.

Triton has an ionosphere made up of electrons and cations (N^+ and probably H^+ and C^+). The ionosphere extends from altitudes of ~100 km to over 800 km. According to one study, the electron density reaches 20,000 electrons/cm^3 at an altitude of 350 km. Although this is a high density, it is still 100,000 times lower than the density of neutral atoms at this altitude.

The tropopause is 8 km above the surface and its temperature may be ~1 K lower than at ground level. One group computed a tropopause temperature of 37 K based on Voyager 2 data. As in the case of Uranus and Neptune, the tropopause is the area with the lowest temperature. Below the tropopause lies the troposphere. Triton's atmosphere is so thin that it may not have a stratosphere.

The lower levels of Triton's atmosphere are illustrated in Figure 2.27. The temperature may increase a little as one gets closer to the surface. The main gas in the troposphere is molecular nitrogen. Small amounts of methane are also present. There are probably trace amounts of carbon monoxide and argon in the troposphere. Other compounds, such as ammonia, water and carbon dioxide are not present because of the low temperatures. These compounds would freeze quickly in Triton's frigid atmosphere.

Why is nitrogen the dominant gas in Triton's atmosphere? There are two reasons: (1) nitrogen ice is abundant on Triton, and (2) this element has a high vapor pressure

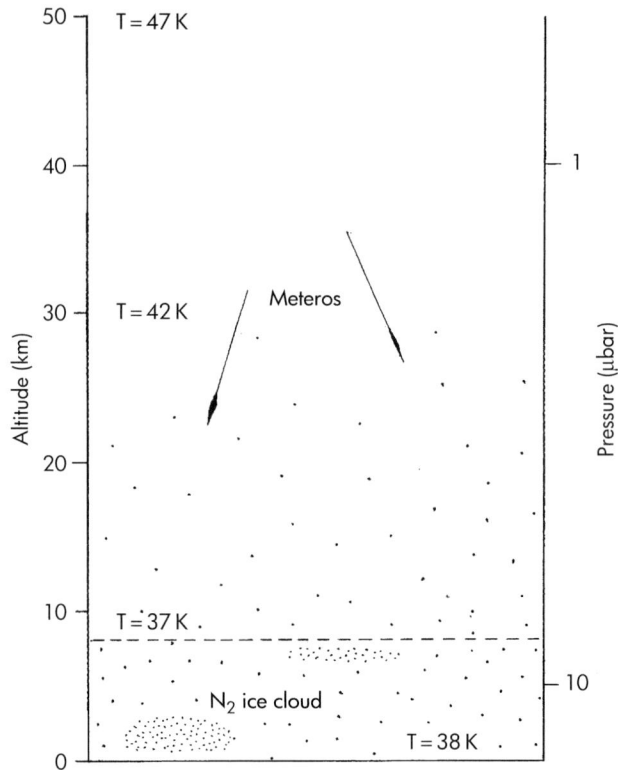

Figure 2.27. A diagram of Triton's lower atmosphere. As in Figure 2.26, the altitudes and atmospheric pressures are shown on the left and right sides, respectively. The dashed line is the tropopause. (Credit: Richard W. Schmude, Jr.)

at 38 K. The thickness of Triton's atmosphere depends on the surface temperature. If the temperature rises to 40 K, the atmospheric pressure at the surface will rise to ~50 μbar. This is more than three times the pressure for a temperature of 38 K. The thickness of Triton's atmosphere changes with the surface temperature and its complex seasons. There is some evidence that its atmospheric pressure changed in the 1990s. In 1989, the pressure was ~14 μbar at the surface, but one group of astronomers observed a stellar occultation by Triton's atmosphere in 1997 and reported that their data was consistent with a surface pressure of ~19 μbar.

Triton's atmosphere is thick enough to cause fast moving meteoroids to burn up. The speed of these objects depends on their relative speed with respect to Triton along with the gravitational acceleration caused by both Neptune and Triton. Fast moving particles with a mass of 0.01 to 1 gram will burn up in Triton's atmosphere and will appear as bright meteors from the surface. One group of astronomers reports that the fast moving meteors burn up at an altitude of ~30 km. Some meteors, however, may be bright until they reach the surface. The thin atmosphere probably slows down lots of micrometeorites, meaning that fewer small craters form on Triton than on moons with no atmosphere. It may also cause larger objects to break apart before reaching the surface.

Winds develop in Triton's atmosphere. One cause of these winds is the unequal distribution of sunlight across the day and night sides of that moon. One group reports that the prevailing winds blow towards the east-northeast direction near the surface between 10°S and 50°S. In 1989, winds at the tropopause were blowing towards the west which was the opposite direction of the prevailing winds at the surface. Stellar occultation data are consistent with high speed winds at an altitude of 100 km.

Hazes, Clouds, and Dust Material from Geysers

Triton's atmosphere contains three types of condensed particles, namely, hydrocarbon haze, clouds and dark material from geysers.

Hydrocarbon haze particles probably lie at all altitudes from the surface up to ~30 km. The particles are made up of hydrocarbon compounds along with compounds containing both carbon and nitrogen. One group of astronomers reports that the optical depth of this haze is less than 0.05 in visible light. Hence, it absorbs very little light. Small haze particles form when sunlight breaks down methane much like what it does in the atmospheres of Uranus and Neptune. Larger hydrocarbons like ethane (C_2H_6) form and condense in Triton's cold atmosphere and begin settling. Haze particles may grow as they settle.

Thin clouds develop at altitudes of between 0 and ~8 km. They may form when the atmospheric temperature drops causing the nitrogen to condense onto suspended aerosols, forming micron-sized ice particles. These clouds can absorb 10 to 20 % of the light falling on them. They can be up to several hundred kilometers long. A very light snowfall (of microscopic snowflakes) may even take place on Triton; however, it is too cold for liquid nitrogen to form in the atmosphere. One group of astronomers detected one haze free area in Triton's atmosphere in 1989. They suggested that a recent snowfall may have cleaned out this part of the atmosphere.

Geysers are the source of a third type of solid particles in Triton's atmosphere and Voyager 2 images have revealed much information about these outbursts. Three images show at least four active geysers. Two of these were named Hili and Mahilani. All four geysers were near 50°S and, hence, may have been caused by changes in solar energy reaching the surface. Alternatively, they may be the result of the release of heat from the interior to the surface. We know that Triton's interior releases some heat as a result of the decay of radioactive elements or heat left over from earlier events. Each of the geysers were between 40 and 1,000 meters across, and material from them rose to an altitude of ~8 km. See Figure 2.28. Hili

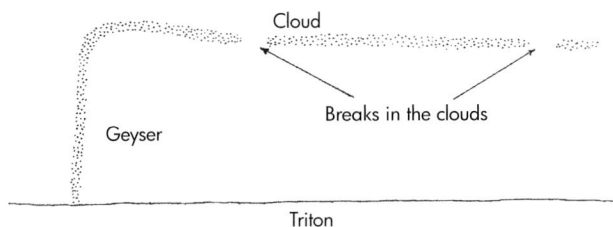

Figure 2.28. A side view of one of Triton's geysers. A jet of material shoots up in a vertical direction until it hits an altitude of 9 km. At this point, it drops back down to an altitude of 8 km, and winds stretch it out into a long cloud. Breaks in the cloud occur because the geyser is not continuous but erupts in spurts. (Credit: Richard W. Schmude, Jr.)

and Mahilani are probably jets of material shooting up from the surface. Each of the geysers ejected ~20 kg of dark aerosols along with a few hundred kilograms of gas each second. The trailing cloud of the Mahilani geyser was caused by winds blowing the material away. One group of astronomers reports a wind speed of ~15 m/s (34 miles/hour) near the tropopause based on the growth rate of the Mahilani cloud. This cloud had several breaks in it, which is consistent with several 10 to 20 minute bursts of activity separated by brief periods of no activity. If the cloud was made up of large particles, they would begin to settle and the cloud altitude would drop; however, this was not observed. Typical cloud particles are less than 5 μm across, and they fall at rates of less than 0.1 meter/second (0.2 miles/hour).

Geyser clouds are faint in Voyager 2 images because they have a low particle density and a low optical depth. One group of astronomers reports that one geyser cloud and its shadow were only a few percent darker than the surface.

How do the geysers form? Figure 2.29 shows how a geyser may develop on Triton. In the first step, an area of sub-surface ice with dark particles heats up by ~3 K. This heat may come from sunlight. Since dark areas absorb more sunlight than brighter ones, they heat up quicker. The overlying ice can also act as a greenhouse by allowing sunlight into a deep layer while trapping the heat given off by the warmer, deep layer. At some point, the pressure builds up. The ice above the high pressure area weakens. This could happen as a result of a phase transition in the nitrogen ice layer, a meteoroid impact or some other mechanism or event. After the overlying layer cracks, the gas and aerosols are forced to the surface and shoot into the thin atmosphere.

There are over 100 dark spots in the southern hemisphere, and almost 90% of them lie between 10°S and 50°S. They may be remnants of geysers. These spots have a slightly reddish color and range in size from ~5 km to over 100 km long. Most of them have sharp southwestern borders but faint northeastern borders. One group of astronomers interprets this as caused by winds blowing in an east-northeast direction.

Triton's Surface

All of our knowledge of Triton's surface features comes from Voyager 2 images. Due to the fact that Voyager 2 did not orbit this moon but instead flew beyond, it imaged less than half of Triton's surface. The best images have resolutions

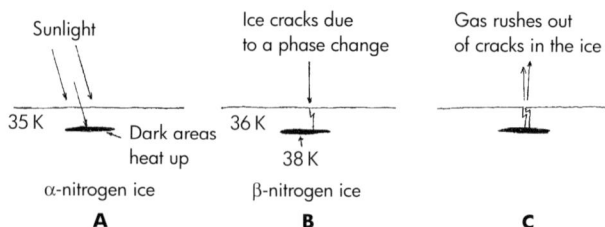

Figure 2.29. A geyser may form when sunlight heats up a darker layer of ice buried below transparent ice. As the darker ice heats up, pressure builds up (Frame A). At that point, the top layer of ice cracks due to a phase change from α-nitrogen to β-nitrogen (Frame B). Finally, gas escapes through the crack (Frame C). (Credit: Richard W. Schmude, Jr.)

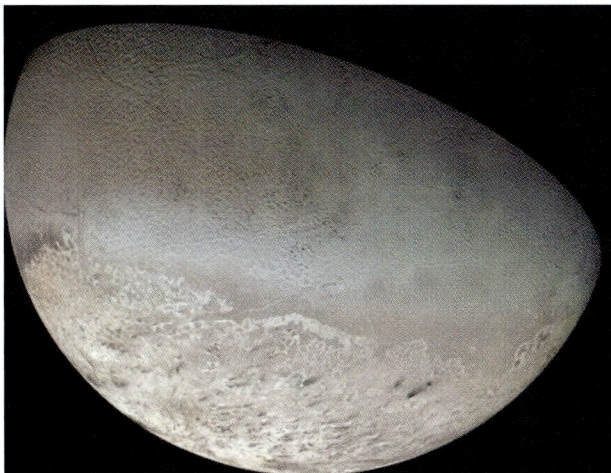

Figure 2.30. Color mosaic of several images of Triton made by Voyager 2. (Credit: Courtesy NASA/JPL-Caltech.)

of ~1 km per pixel; hence, scientists are unable to resolve features smaller than 2 km in the best images. The following discussion is based on only the areas imaged.

There are several topographical features on Triton larger than 1 to 2 km. Several of these are visible in the Voyager 2 image in Figure 2.30 and in the map in Figure 2.31. I will describe several types of features on this moon.

Triton has a low density of craters, which may be due to cryovolcanic activity. Essentially, this is a process where low melting point materials like water and ammonia are released to the surface. At that point, these materials bury the pre-existing craters and solidify later. Alternatively, Triton's surface may have heated to the point where the ices melted, causing all previous topographic features to disappear. One event that could trigger widespread cryovolcanic activity or surface melting is Neptune's capture of Triton. The resulting tidal forces from Neptune could have caused Triton's shape to undergo small, periodic changes. This, in turn, would cause internal friction, which would lead to a build-up of internal heat. For instance, we know that Jupiter's moon, Io, has active volcanoes because of tidal forces.

There are several craters on Triton that have diameters between 2 and 27 km. Craters with diameters less than 2 km are undoubtedly present, but were not resolved in Voyager 2 images. Complex craters are present, and a few of them have rim ejecta. Mazomba is the largest crater imaged, with a diameter of 27 km, and is shown in Figs. 2.30 and 2.31. This crater, along with a few of the larger ones, have central peaks and are complex. The transition diameter between simple and complex craters is ~11 km on Triton, which is lower than the transition size on Uranus' moon Ariel (15 km). This difference is probably due to Triton's higher gravity. The depth-to-diameter ratio for craters on Triton is around 0.1, which is similar to the craters on the large moons of Uranus.

There is some evidence that the crater distribution on Triton is different from that on Miranda and Earth's Moon. In short, there are more small craters for each large crater on Triton than on Miranda. The different distribution on Triton may be due to its thin atmosphere or due to different types of impacting bodies hitting it compared to moons closer to the Sun.

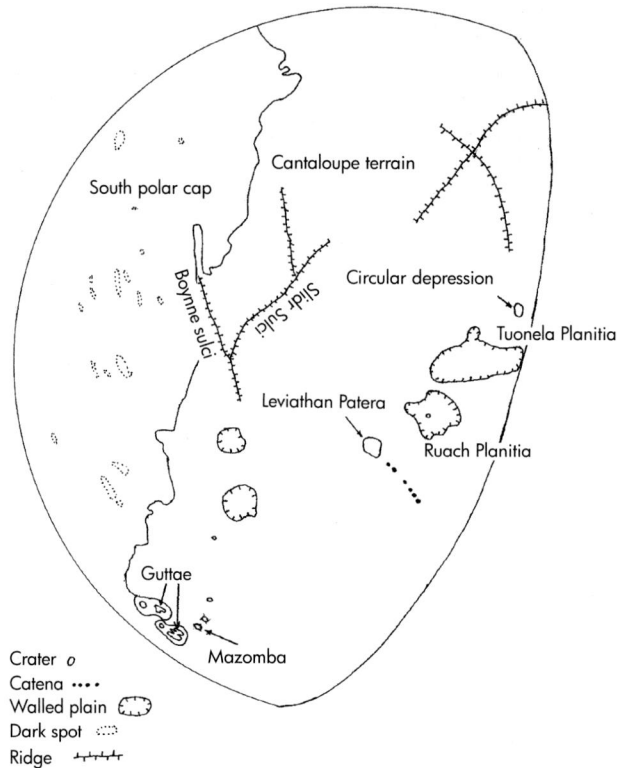

Figure 2.31. A map of Triton showing a few of the larger features on it. (Credit: Richard W. Schmude, Jr.)

Triton's leading hemisphere has a higher crater density than its following hemisphere. This is consistent with its sweeping up of particles as it moves around Neptune. Essentially meteoroids have to catch up to Triton to strike its following hemisphere, whereas this is not the case for objects striking its leading hemisphere. In addition, rocks striking Triton's leading side will usually have higher impact speeds, thereby creating larger craters than those striking the other side. Earth behaves in a similar way. Sporadic meteors and most shower meteors are more numerous in the two hours before sunrise than during the two hours after sunset.

All of Triton's craters have a fresh appearance. This is due to its atmosphere (which slows down incoming meteoroids) and the low amount of erosion on Triton. As a result, there is little meteoroid erosion. The atmosphere has winds, but they exert very little force on the surface due to low gas density; therefore, wind erosion is also minimal.

The fresh appearance of the craters is also due to the rigid surface materials. We know that some solid nitrogen ice exists on Triton; however, one group of astronomers points out that a crater in a two-kilometer layer of nitrogen ice will sag over a long period of time. This group suggests that large amounts of water ice are near the surface and, since it is more rigid, craters retain more topographical relief in this material. Hence, the fresh appearance of the craters is consistent with large amounts of rigid water ice near Triton's surface.

Triton has circular depressions in a wide range of sizes. These are probably not craters. These features often lack raised rims and have a different size distribution from craters. Many small depressions are on top of their own 100 to 200 meter tall hill, which suggests some kind of volcanic origin. They range in diameter from ∼4 to ∼100 km. Many of the smaller pits may be the result of escaping gases from the interior. One of the largest circular features is Leviathan Patera which is shown in Figures 2.30 and 2.31. A few smaller pits are arranged like catenas (or chains of craters).

A third type of feature on Triton is the ridge. A typical ridge is ∼15 km wide and is up to 1,000 km long. One ridge, Slidr Sulci, rises ∼200 meters above the surface and its slopes are gentle and similar to that of a wheelchair ramp. A second ridge, Boynne Sulci, starts in the south polar cap and merges with Slidr Sulci forming a sideways "Y" pattern. Slidr Sulci lies above an older ridge, which is evidence that the ridges formed at different times. Figures 2.30 and 2.31 show several ridges.

The fossae are long, narrow trenches. Two of these features Yenisey Fossa and Jumma Fossa are ∼2 km wide but are each over 200 km long and they are almost straight. Each of the fossae cuts across at least two types of terrain and is ∼100 m deep. An episode of expansion of Triton may have created these features.

The Guttae are unique to Triton. They appear as darker areas that are surrounded by a bright border or aureole. See Figure 2.32. Akupara Maculae and Zin Maculae are two examples of Guttae. The aurolae are less than 100 m thick and are probably at least several meters above the darker portion of the Guttae. They reflect 90% of the light falling on them compared to 70% for the darker inner region. The darker areas are probably nitrogen ice with some methane and other impurities. Figs. 2.30 and 2.31 show several Guttae near the edge of Triton.

Figures 2.30 and 2.31 show Ruach and Tuonela Planitia. These features are examples of walled plains. They are large flat areas surrounded by walls ∼100 m to ∼200 m high. The Planitia are smooth areas, except for their pits and an occasional crater. Ruach Planitia is flat to within 100 m. At least three of these features have clusters of pits near their centers. These plains may be areas where large amounts of slushy ice made its way to the surface.

The oldest area on Triton is the "cantaloupe" terrain. We know that it is old because other features cut across it. This area resembles the surface of a cantaloupe–hence the name. There are dozens of pits and ridges covering the surface. The pits are ∼20 km across and may be due to the release of gas underneath, resulting in the collapse of the top layer of ice. Figure 2.33 shows the cantaloupe terrain.

Most of Triton's surface reflects about as much visible light as clean, fresh snow. We know this from both Voyager 2 images and Earth-based brightness measurements. Most of the areas reflect about 70 to 80 % of the light falling on them, which is about two to three times what the large moons of Uranus reflect. This may be due to smaller amounts of rock and carbon material on Triton's surface, or it may be due to the fact that fresh nitrogen snow is deposited on Triton from its atmosphere. There are several small spots on Triton with sizes of a few kilometers that reflect only about 25 % of the visible light falling on them. These may be areas of exposed rock or carbon rich material mixed with ice.

Most of Triton's surface is covered with nitrogen ice with small amounts of methane and carbon monoxide impurities. The frozen nitrogen is at least several centimeters thick. Much of this ice may be in the form of centimeter sized ice chunks. The nitrogen layer may lie on top of a thick layer of water ice or water ice with other dissolved materials. See Figure 2.34. There are no large areas of pure

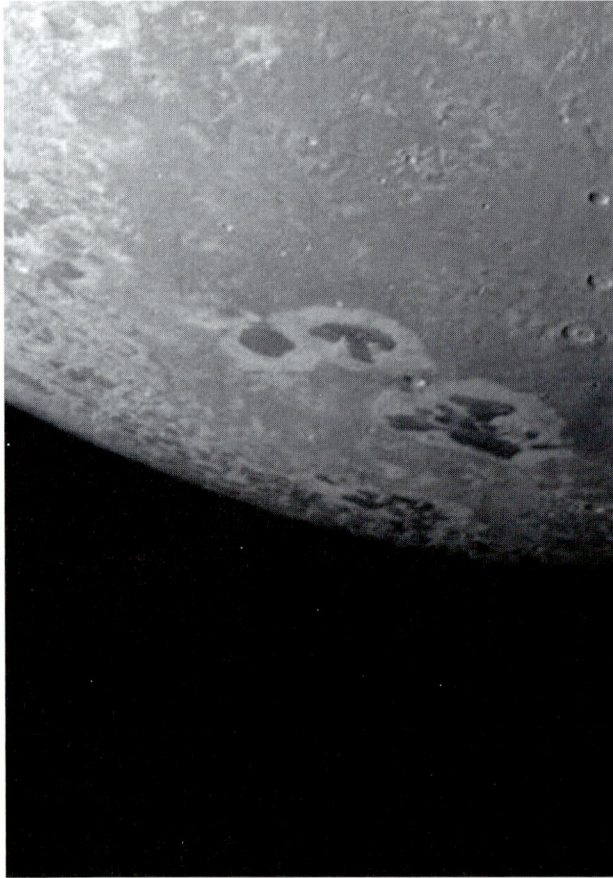

Figure 2.32. A color image of the Guttae on Triton along with a few craters. This figure covers an area 400 km across. The largest crater in this image is Mazomba which is 27 km across. (Credit: Courtesy NASA/JPL-Caltech.)

Figure 2.33. A color image of the cantaloupe terrain (center) and of a portion of Triton's south polar cap (left). Astronomers used several Voyager 2 images to construct this picture by combining violet, green and ultraviolet images. (Credit: Courtesy NASA/JPL-Caltech.)

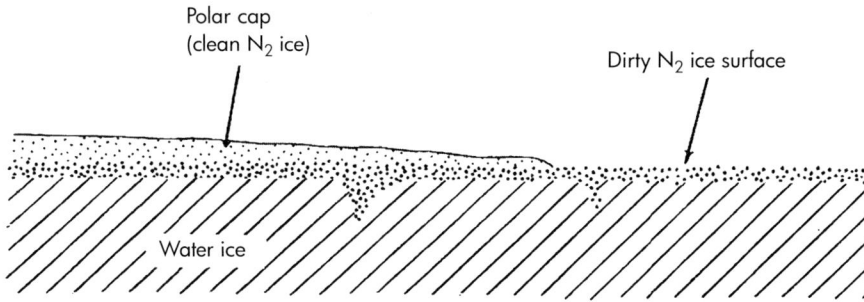

Figure 2.34. A cross-section of Triton's surface. The bright polar ices are fresh and are brighter than the underlying layer of permanent nitrogen ice. Below this layer lies a rigid layer of water ice. (Credit: Richard W. Schmude, Jr.)

frozen carbon monoxide (CO) since CO is not abundant in the atmosphere. Voyager 2 failed to detect any gaseous carbon monoxide. Water ice (in a crystalline state) is also on the surface. Small amounts of ethane are present as well. The source of this material is either atmospheric haze that has settled on the surface or solid state reactions involving solar ultraviolet light and methane on Triton's icy surface.

One group of astronomers searched for specular (or mirror-like) reflection on Triton but was unable to find it. Smooth or glazed surfaces can produce this type of reflection. Any ices on Triton with these characteristics would be less than a few kilometers across. In 2001, a group of amateur and professional astronomers discovered an example of specular reflection on Mars.

At low temperatures, nitrogen can exist in either the alpha (α) or beta (β) phase. The atoms are arranged in a high-density cubic structure in the α phase, whereas they are arranged in a low density hexagonal structure in the β phase. See Figure 2.35. At temperatures below 35.6 K, the α phase is more stable, whereas the β phase is more stable at higher temperatures. If ice is at 35 K and is heated to 36 K, a phase transition would take place. Since there is a difference in density, pressure will build up in the ice and it may crack. Low amounts of impurities (less than 1%) will have little effect on the α to β phase transition temperature. Most of the nitrogen ice on Triton is in the β phase; however, the temperature may fall low enough for the α phase to develop.

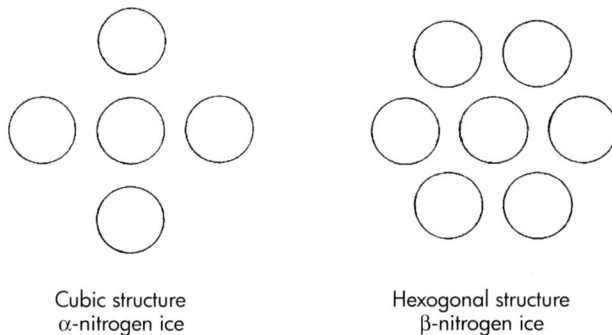

Cubic structure
α-nitrogen ice

Hexogonal structure
β-nitrogen ice

Figure 2.35. Arrangement of the high density alpha (α) phase of nitrogen (left) and the lower density beta (β) phase (right). Each circle represents a nitrogen atom. (Credit: Richard W. Schmude, Jr.)

Triton's south polar cap reflects ~80% of the light falling on it, which is a few percent more than the rest of the surface. The polar cap is made up of nitrogen ice containing impurities. The higher albedo of the polar cap suggests that there are fewer impurities in the polar ice than in the surrounding ice. The cap is probably thickest near Triton's south pole and thins out towards the equator. One group of astronomers estimates an upper limit to its thickness of 1 to 2 km. The polar cap extends from the pole to around 10°S on the side facing Neptune. (Keep in mind that Voyager 2 imaged primarily the side of Triton facing Neptune.) It has a slightly pinkish color, which may be due to small amounts of hydrocarbon impurities.

How can Triton's south polar cap extend down to latitudes near the equator? One explanation is a combination of Triton's low temperature along with its bizarre seasons. During the early 20th century, Triton's southern hemisphere experienced a very cold winter. During the 1989 Voyager encounter, temperatures in Triton's southern hemisphere were undoubtedly rising, but the effects of the cold winter may have been left behind in the form of a large polar cap.

Haze may also play some role in the size of Triton's south polar cap. A haze layer will block out some sunlight, causing lower surface temperatures. A thin haze was present above the polar cap in 1989 and, hence, this may have caused lower temperatures on the surface.

Small amounts of rocky material are also on the surface. We know that rocky material does not cover a large percentage of the surface because of Triton's high albedo. Some of it may be olivine and iron (II) sulfide. The rocky material may have come from meteorites reaching Triton's surface or may have been extruded from the interior.

Triton's Color and Brightness

Since 1952, astronomers have measured Triton's color and brightness. Table 2.8 includes some of Triton's photometric constants. Triton's average B–V, V–R and R–I color indexes are: 0.73 ± 0.03, 0.39 ± 0.03 and 0.38 ± 0.02 and its Bond albedo is 0.85 ± 0.05. Unlike Uranus' large moons, Triton lacks an opposition surge. This is due probably to the nature of its surface, which is mostly nitrogen ice instead of rock. Triton is a few percent brighter at western elongation than at eastern elongation. This brightness difference is larger in visible light than in near-infrared light. There is some evidence that the brightness difference between the two hemispheres dropped between 1950 and 2000. This may be due to the formation of new ice layers.

There is some evidence that Triton's B–V color index changes. Its B–V values in 1952, 1977–78, 1989 and 1997–2000 were 0.77, 0.72, 0.70 and 0.73, respectively. This change may be due to the changing position of the sub-Earth latitude. In the early 1950s, the sub-Earth latitude was near the equator, but by 2000, it was near 52°S.

Triton's Interior and Past History

What is Triton's interior like? Triton's density is at least twice that of water ice or other ices. Therefore, a large amount of rock must be present in its interior.

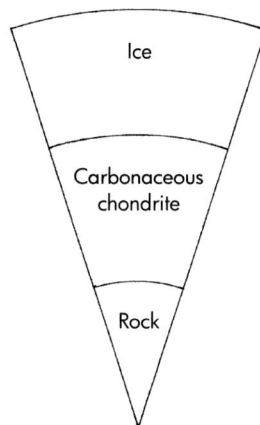

Figure 2.36. A cross-sectional view of Triton's interior. (Credit: Richard W. Schmude, Jr.)

Figure 2.36 shows a possible cross-section view of Triton. A small rocky core may lie at the center, which is surrounded by a thick layer of carbonaceous chondrite material and a top layer of ice. The carbonaceous chondrite layer may contain lots of dark carbon material similar to what is found in carbonaceous chondrite meteorites. Triton releases a small amount of internal energy due to the decay of radioactive elements. Additional heat may come from heat left over from Triton's formation or from heat generated after it was captured by Neptune.

Does Triton have liquid water in its interior? Internal heat may prevent some deep water from dropping to the freezing point and, hence, Triton may have a liquid layer in its interior. One group of astronomers points out that a liquid "ocean" layer may exist 20 to 140 km below the surface. If so, Triton's top layer of ice would serve as insulation. We know that there are layers of liquid water below the ice cap of Antarctica on Earth and, hence, Triton may have an insulated deep layer of liquid water which may be mixed with ammonia or other materials.

What is Triton's geological history? A possible scenario is that, after formation; it experienced a heavy bombardment of meteoroids. Triton may have formed as either a single or a binary object. It made a close approach to Neptune and was captured. It initially had a highly elliptical orbit but, after awhile, tidal forces forced it into a circular orbit. During this time, large amounts of cryo-volcanic activity took place due to tidal heating. This volcanic activity erased all pre-existing features. Then cantaloupe-like terrain developed. Finally, other features like craters, ridges and walled plains developed. Lastly, the polar cap and many dark features developed. Much of the south polar cap may be just a few centuries old.

Collision Fragments

The five innermost moons – Naiad, Thalassa, Despina, Galatea and Larissa – have revolution periods that are shorter than Neptune's rotation period. Tables 2.7 and 2.8 summarize their characteristics. The estimated masses are based on an average density of $1.5 \, g/cm^3$. I will discuss four of these moons.

Larissa

Larissa was first detected during a stellar occultation on May 24, 1981. Astronomers recorded the occultation using two different telescopes near Tucson, Arizona. They reported that this satellite was 50,000 km from the cloud tops of Neptune, and that it was at least 180 km across. The discoverers predicted that this moon was so close to Neptune that it would be beyond the detection limit of cameras in 1981. Scientists on the Voyager 2 team confirmed the presence of Larissa and were the first to refine accurately its orbit.

Figure 2.37 shows a Voyager 2 image of Larissa. This moon has a round edge. There appears to be a 40 km crater near the limb. There are also other irregularities, but they are difficult to make out because of the low resolution of the image.

Despina and Naiad

Despina and Naiad have shapes similar to potatoes. If these moons have synchronous rotations then they will change in brightness due to geometry. Despina will change by ∼0.15 magnitudes and Naiad will change by ∼0.3 magnitudes. Galateo, Larissa and Proteus have nearly spherical shapes; hence, brightness changes due to a changing geometry should be minimal.

Thalassa

Thalassa has a unique shape among the inner moons. Its two longest dimensions are almost equal. This moon has a shape similar to that of a hamburger. As a result of its shape, it probably has almost the same brightness as it revolves around Neptune.

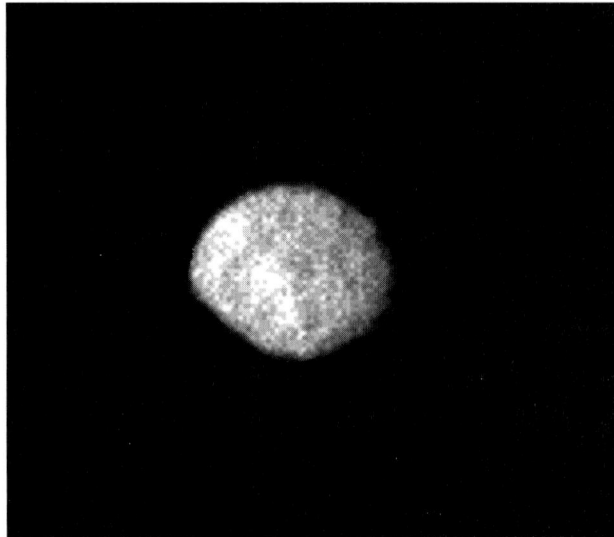

Figure 2.37. Voyager 2 image of Larissa taken on Aug. 24, 1989, at a distance of 2.45 million km. (Credit: NASA and the NSSDC.)

Proteus

Proteus undergoes synchronous rotation, and its orbit lies very close to Neptune's equatorial plain. Figure 2.38 shows a Voyager 2 image of this moon and Figure 2.39 shows a map of part of its surface. As in the case of Triton, Voyager 2 imaged less than half of the surface of Proteus. The area that was imaged is very dark with almost no bright features. A 210 km basin is present. The impact that created this feature almost destroyed Proteus. A smaller 80 km crater lies on top of the basin. The surface contains also several smaller craters.

Proteus is larger than Saturn's moon, Mimas (average radius = 199 km), and is almost as large as Uranus' moon, Miranda (average radius = 236 km). In spite of these similarities, Proteus is not as round as those moons. This difference may be due to a different composition of this moon than of both Mimas and Miranda. The latter two moons have albedos that are more than a factor of four higher than Proteus. Perhaps Proteus has a lower percentage of ices and a higher percentage of

Figure 2.38. Voyager 2 image of Proteus taken on Aug. 25, 1989, at a distance of 144,000 km. (Credit: NASA and the National Space Science Data Center.)

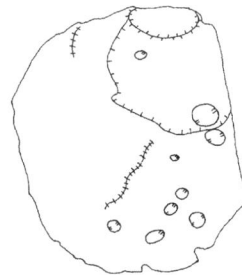

⊖ Crater
⤛⤛⤛ Trench
⤛⤛⤛ Raised rim
⤛⤛⤛ Scarp

Figure 2.39. Map of Proteus. (Credit: Richard W. Schmude, Jr.)

more rigid and dark material than the other two moons. Since Proteus does not have a round shape, it is considered to be a collision fragment instead of a "regular" moon.

Proteus reflects a little more green than blue light, which is different than Miranda. Proteus' albedos for the J, H and K filters are 0.026, 0.028 and 0.047, respectively. These albedos are a factor of 5 to 7 lower than the corresponding values for Miranda. The difference in albedos is further evidence that Proteus and Miranda are covered with different material.

Captured Objects

The six moons Nereid, Halimede, Sao, Laomedeia, Neso and Psamathe which are far from Neptune, are considered captured objects. Three of them (Halimede, Neso and Psamathe) have retrograde orbits. Psamathe and Neso are farther from their primary than any other moon in the Solar System. These two make one trip around Neptune every 24+ years, which is longer than for any other known moon in our Solar System. All six outer moons are assumed to have the same albedo as Nereid. Estimated sizes and masses are based on this assumption. Brief discussions of two of these moons follow.

Nereid

Voyager 2 was unable to fly close to Nereid and as a result, we do not have any close-up images of it. We do not know its shape, or the types of surface geological features or its rotation rate. Since Nereid is so far from Neptune, there is a good chance that it has chaotic rotation. We are fairly confident of its size from distant Voyager images. Most of our knowledge of this moon comes from Earth-based studies.

Two astronomers carried out brightness measurements of Nereid over a twenty year period, and they reported that it can brighten by a factor of five in just a few days. Much of this change is due to bright and dark areas coming into view. Specular (or mirror-like) reflection may also cause Nereid to brighten. Its average color indexes are: B–V = 0.69, V–R = 0.43 and V–I = 0.72 and V–K = 1.6. The first three values are similar to those of Triton.

What is Nereid's surface like? Some clues to the question come from spectroscopic studies carried out with large telescopes. Recently, one group of astronomers detected water ice frost on it. Since Nereid's albedo is around 0.26 at a phase angle of $\sim 0°$ (which is much lower than that of pure ice), other materials are present on it. There is a small possibility that water from Nereid's interior is released similarly to what happens on Saturn's moon, Enceladus. Since Nereid lacks a significant atmosphere, there are undoubtedly craters on its cold surface. In addition, there are probably many rocks and smaller chunks of ice on its surface.

Halimede

This moon probably has an irregular shape because its brightness changes by at least 0.7 magnitudes as it moves around Neptune. One group of astronomers used

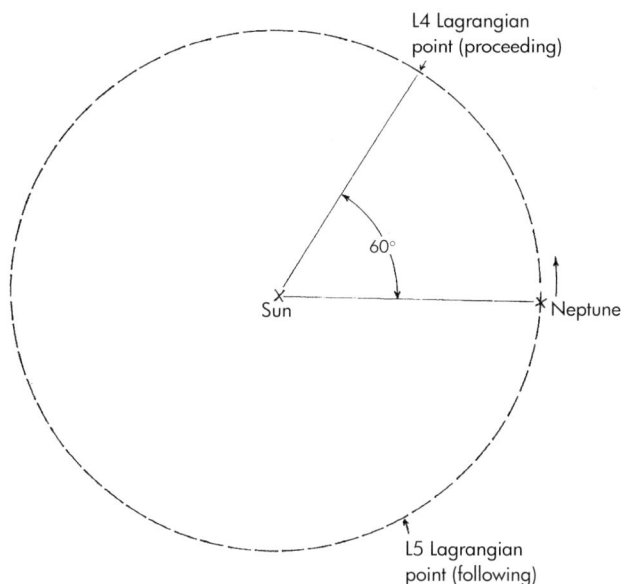

Figure 2.40. The dashed circle represents Neptune's orbit. If Neptune is at the X, the two Lagrangian points are at L4 (proceeding) and L5 (following) positions. These two points are areas of extra stability. Several Trojan objects lie near Neptune's L4 point. (Credit: Richard W. Schmude, Jr.)

the Magellan and Keck II telescopes to measure its color indexes as B–V = 0.70 and V–R = 0.37. These values are similar to those of the Sun, which is consistent with that moon having a color similar to that of sunlight. Halimede's B–V and V–R values are also close to those of Triton and Nereid.

Trojan Asteroids

A Trojan asteroid is an object that orbits either ahead of or behind a planet. The planets Mars, Jupiter and Neptune all have Trojan asteroids. As of March 2007, Neptune had five known Trojan asteroids. These asteroids have the same orbital period as Neptune and intersect Neptune's orbit at the L4 Lagrangian point. This is an area of extra stability. See Figure 2.40. These objects are believed to reflect just a few percent of the sunlight falling on them and have diameters of 100 to 200 km. Because they are 60° ahead of Neptune, they reach opposition two months after it. There are currently no known objects that lie near the L5 Lagrangian point; part of this may be due to the fact that in the early 21st century, it lies near the rich star field in Sagittarius.

Chapter 3

Pluto and Its Moons

Introduction

Pluto has been a mysterious object since its discovery in 1930. It is too small to show a disc in all but the largest telescopes. No probe has visited it and, hence, most of what we know about its physical characteristics is from telescopic data. Recent Hubble Space Telescope images show a few crudely resolved surface features on this distant world. Position measurements along with mathematics and gravitational theory have yielded information on Pluto's orbit. The New Horizons probe launched in 2006 will not reach Pluto until 2015 and, hopefully, we will get our first close-up images of its surface.

Up until 2006, Pluto was classified as a planet. For many years, it was thought to be similar in size to Mars and Mercury. This changed in the late 1970s when Pluto's largest moon, Charon, was discovered. Astronomers measured both the orbital period of Charon and its distance from Pluto, and from this, computed the combined mass of Pluto and Charon. They found that the combined mass was less than that of our Moon. The small size of Pluto, along with the discovery of similar sized objects in our outer Solar System, forced astronomers to make a decision about what is and is not a planet.

In 2006, hundreds of astronomers making up the International Astronomical Union (IAU) met in Prague, Czech Republic, and, after long debates, defined not only "planet" but a new class of objects called "dwarf planets". Table 3.1 summarizes characteristics of a planet and a dwarf planet. While Pluto meets two of the characteristics of a planet, it failed to meet the final characteristic, namely, that it has cleared its neighborhood around its orbit. Hence, it is not a planet under the new definition. It meets the criteria of a dwarf planet and, hence, this is how it is classified.

In this chapter, I would like to first discuss Pluto's orbit and seasons since these will affect its atmosphere. I will then discuss Pluto's atmosphere, interior, magnetic environment, surface and moons.

Orbit and Seasons

Table 3.2 lists current orbital elements of Pluto. Figure 3.1 shows the orbits of Uranus, Neptune and Pluto. As it turns out, Pluto can be a little closer to the Sun than Neptune. This is because of the relatively high eccentricity of Pluto's orbit. When Pluto is near perihelion, it is much closer to the Sun than when it is at aphelion. Pluto reached perihelion in 1989 and was closer to the Sun than Neptune

R.W. Schmude, Jr., *Uranus, Neptune, and Pluto and How to Observe Them*,
DOI: 10.1007/978-0-387-76602-7_3, © Springer Science+Business Media, LLC 2008

Table 3.1. Characteristics of a planet and a dwarf planet as defined by the International Astronomical Union in 2006

A planet is a celestial body that:
    (a) is in orbit around the Sun
    (b) has sufficient mass for its self gravity to overcome rigid body forces so that it assumes a hydrostatic equilibrium (nearly round) shape; and
    (c) has cleared the neighborhood around its orbit.
A dwarf planet is a celestial body that:
    (a) is in orbit around the Sun;
    (b) has sufficient mass for its self gravity to overcome rigid body forces so that it assumes a hydrostatic equilibrium (nearly round) shape;
    (c) has not cleared the neighborhood around its orbit; and
    (d) is not a satellite.

Table 3.2. Characteristics of Pluto

Characteristic	Value
Radius	1,160 km
Surface area	$1.77 \times 10^7 \, km^2$
Mass	$1.305 \times 10^{22} \, kg$
Density	$2.0 \, g/cm^3$
Surface composition	Mostly nitrogen ice mixed with methane impurities, tholins, and carbon monoxide; possibly pure methane, water ice and ethane
Period of rotation	6.387 days
Period of revolution (average)	248 years
Inclination of equator to orbit	57.5°
Average distance from Sun	39.5 au
Average opposition distance from Earth	38.5 au
Orbital inclination	17°
Orbital eccentricity	0.25
Vo	15.37
V(1,0)	−0.54
Average geometric albedo (V filter)	0.55

for about two decades in the late 20th century. This will not occur again until the 23rd century. The dashed portion of Pluto's orbit in the figure shows when that planet lies north of the ecliptic, and the solid portion shows when it is south of the ecliptic. Since Neptune's orbit lies nearly in the ecliptic, Pluto will often lie far to the north of it. As it turns out, when Pluto is closest to the Sun, it is also well north of Neptune's orbital plane and, hence, there is no chance of it colliding with that planet.

    The four giant planets – Jupiter, Saturn, Uranus and Neptune – exert gravitational tugs on Pluto. These tugs are called perturbations. Neptune, which is closest to Pluto, exerts strong perturbations on it, and, as a result, the orbital elements of

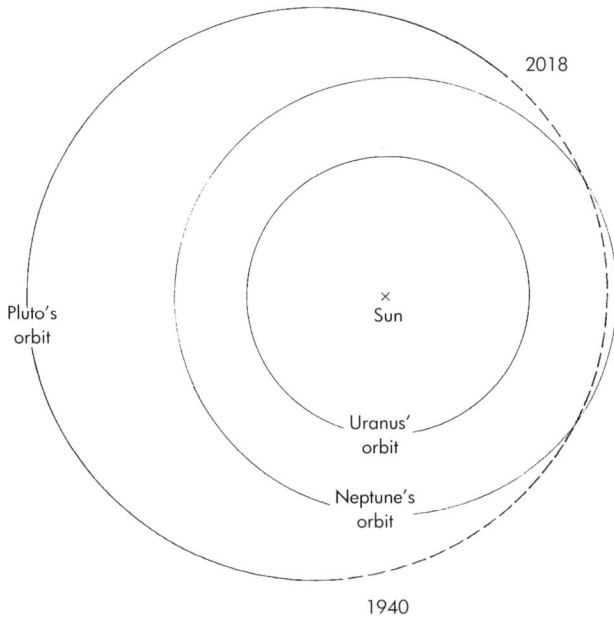

Figure 3.1. The orbits of Uranus, Neptune and Pluto are shown here. The dashed curve shows the section of Pluto's orbit that is north of the ecliptic. Pluto crossed the ecliptic in about 1940 and it will cross it again in late 2018. Since the orbits of Uranus and Neptune have such low inclinations, they lie close to the ecliptic plane. Pluto was closer to the Sun between 1979 and 1999 than Neptune, and this is shown in the Figure. (Credit: Richard W. Schmude, Jr.)

Pluto change over time. As an example, Pluto's orbital period can be as short as ~244 years and as long as ~252 years. The average orbital period is 248 years. Computing changes in Pluto's orbit is a complex task and requires knowledge of the changes in the orbits of the four giant planets. High-speed computers have allowed people to compute changes in Pluto's orbit well into the future and into the distant past. These studies have yielded two important findings: (1) Pluto's orbit will remain stable for at least the next 4000 million years and (2) Pluto is probably not an escaped moon of Neptune.

How can Pluto avoid colliding with Neptune? The answer is that Pluto is in a 3-to-2 resonance with that planet; this means that, for every two trips around the Sun that Pluto makes, Neptune makes three. This stabilizes Pluto's orbit. As mentioned earlier, one must remember also that Pluto lies north of Neptune's orbital plane when it makes its closest approach to the Sun. Because of its orbit and resonance with Neptune, Pluto never gets closer than 17 au from that planet; in fact, it gets closer than this to Uranus. Figure 3.2 show the locations of Neptune and Pluto in 1840, 1920 and 2005.

Pluto is probably not an escaped moon of Neptune, because of its 3-to-2 orbital resonance with that planet. One group suggests that instead of forming near Neptune, it probably formed in a nearly circular orbit beyond Neptune. During this time, Pluto was always further from the Sun than that planet. Later on, Neptune's orbit expanded as a result of gravitational perturbations from the other three giant planets. At some point, Neptune's orbit reached the point where it was at a 3-to-2 resonance with Pluto. This resonance changed Pluto's orbit into one with a higher eccentricity.

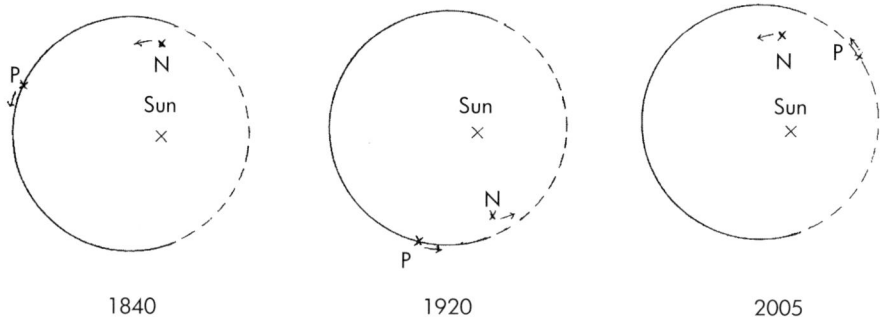

Figure 3.2. Due to the 3-to-2 resonance of Neptune and Pluto, these two bodies will not cross paths for at least the next 4000 million years. The positions of Pluto (P), Neptune (N-small ×) and the Sun (large ×) are shown in the figures for 1840, 1920 and 2005. The dashed line in Pluto's orbit lies north of the ecliptic plane and the solid part lies south of the ecliptic plane. (Credit: Richard W. Schmude, Jr.)

Like Venus and Uranus, Pluto rotates in the opposite direction that it revolves. Pluto's equator is currently tilted at an angle of ~57.5° to its orbit. As a result, the sub-solar latitude reached ~57.5°N in the 1940s and it will reach ~57.5°S in 2029; consequently, at the present time its polar regions receive more sunlight than its equatorial regions over the course of one Pluto revolution. Throughout this chapter, I will use the IAU Convention, which states that Pluto's southern hemisphere is the one that faced the Sun in 1989.

Pluto has seasons. Its northern hemisphere went through autumnal equinox in late 1987 and will reach its winter solstice in mid 2029. Between 1987 and 2029, the southern hemisphere is facing the Sun and Earth. Since Pluto's orbit has a relatively high eccentricity its seasons are not of equal length. For Pluto's northern hemisphere, summer and fall are each only ~42 years long, whereas winter and spring are each ~83 years long. The southern hemisphere has a short spring because Pluto is closest to the Sun at that time; hence it moves faster in its orbit. Temperatures are probably highest in the southern hemisphere during the spring (instead of the summer) because it is so much closer to the Sun then. The seasons for Pluto's northern hemisphere are different. Temperatures in that hemisphere probably reach a maximum value during the summer since Pluto is much closer to the Sun then. Furthermore, in spite of the unequal seasons, both hemispheres get about the same amount of sunlight during a Pluto year. Pluto's atmosphere probably grows during the short, southern spring and probably shrinks during most of the remaining time.

Atmosphere

Pluto has a thin atmosphere, which is probably made up mostly of nitrogen along with small quantities of methane and carbon monoxide. The reason why we are not certain of the composition is because it is very difficult to detect gaseous nitrogen with spectroscopy. We believe that nitrogen is the main constituent in Pluto's atmosphere because it has a significant vapor pressure at Pluto's low temperatures and it is on that body's surface. Figure 3.3 shows a cross-section view of Pluto's atmosphere and Table 3.3 lists characteristics of its atmosphere. One important difference between the atmospheres of Pluto and the Neptunian

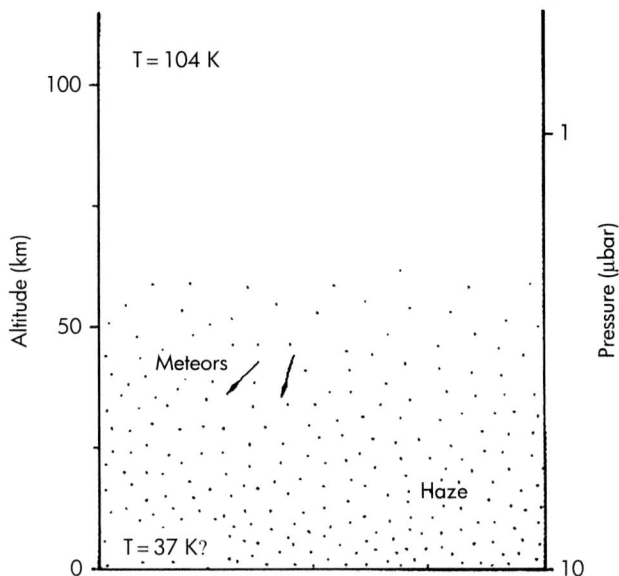

Figure 3.3. A cross-section of Pluto's atmosphere in 2002. Some haze is probably present in its atmosphere and is shown as dots. Fast moving meteorites probably burn up in Pluto's thin atmosphere. The altitude is the distance above Pluto's solid surface. (Credit: Richard W. Schmude, Jr.)

moon Triton is temperature. At the 1.3 μbar level, the atmospheric temperature is ~100 K on Pluto, but is only ~45 K on Triton.

Methane is probably responsible for the higher temperatures in Pluto's atmosphere. Essentially, higher concentrations of this gas mean that more infrared light from the Sun is absorbed by the atmosphere. This will cause the temperature to rise. Since there is a lower concentration of this gas on Triton, the atmospheric temperature remains low. We are not sure why Pluto has more gaseous methane than Triton. One possibility is that it has areas of pure methane ice, whereas all methane ice on Triton is dissolved in nitrogen. Methane that is dissolved in another compound will have a lower vapor pressure than in the pure state for a given temperature. More methane will escape from pure ice than from the solid solution containing methane.

Ethane and hydrocarbon hazes may be present in Pluto's atmosphere. They would form when solar ultraviolet light reacts with methane. This haze may affect

Table 3.3. Characteristics of Pluto's atmosphere	
Characteristic	Value
Composition	Probably mostly Nitrogen with small amounts of methane and carbon monoxide
Surface pressure (1988)	~5 μbar
Surface pressure (2002)	~10 μbar
Atmospheric Temperature (2002)	104 K at an altitude of ~60 km
Total mass	~3 × 10^{13} kg (2002)

Pluto's atmospheric temperature. That body may have a thicker haze layer than Triton due to its higher methane abundance. Occultation data in 1988 are consistent with Pluto's atmosphere being more opaque at lower altitudes than at higher ones. One explanation for this result is that an opaque layer of haze was at the lower altitudes. Astronomers did not measure a similar opacity change during the 2002 and 2006 occultations, which suggests that the haze at lower altitudes became thinner. More occultation data are needed to determine if Pluto has haze layers and if so, how they change with time.

If hazes are present on Pluto, they could affect its brightness and how the brightness changes with longitude. A combination of high resolution images and brightness measurements may shed some light on any such hazes.

Pluto may have geysers like Neptune's moon Triton and Saturn's moon Enceladus. During the early 21st century, parts of Pluto are either reaching or will reach their highest temperatures in 2.5 centuries. This may trigger geysers.

Two changes occur as Pluto moves around the Sun – the Pluto-Sun distance changes and the season changes. Pluto was closest to the Sun in 1989 and, hence, much of its surface was warming up. There is a chance that temperatures in Pluto's southern hemisphere will continue to increase until about 2015. This would be due to the seasonal lag in temperature.

During 1988, 2002, 2006 and 2007, Pluto moved in front of stars, and astronomers recorded how its atmosphere affected the starlight. As it turns out, this data revealed that between 1988 and 2002, Pluto's atmosphere became twice as thick, and that its atmospheric temperature remained nearly the same. The rise in pressure is probably due to a small rise in the surface temperature. Between 2002 and 2007, the atmospheric temperature and pressure remained nearly the same.

Due to its elliptical orbit, Pluto can be as close as 30 au to the Sun at perihelion and as far as 49 au from the Sun at aphelion. This will affect its atmospheric pressure because Pluto receives much more solar radiation per unit time at perihelion than at aphelion. This will likely cause temperature changes, which, in turn, will cause changes in the atmosphere. Essentially the atmosphere is probably thickest when the surface is warmest, and is thinnest when the surface is coolest. One model predicts a surface pressure of 15 µbar for 2015 and a pressure of 0.3 µbar near aphelion, which will occur in the early 22 nd century. If this is correct, Pluto's atmosphere will be near its maximum thickness when the New Horizons probe arrives. The changing thickness of Pluto's atmosphere is shown in Figure 3.4.

We are not sure if the atmosphere changes as a result of rotation. There is a chance that thin clouds develop during Pluto's long and cold nights, but dissipate

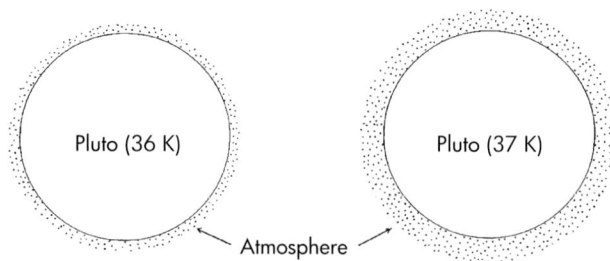

Figure 3.4. The atmosphere (shaded area) becomes thicker when Pluto's surface temperature rises from 36 K to 37 K. The surface temperature may have increased by about one Kelvin between 1988 and 2002, which led to a thicker atmosphere in 2002. (Credit: Richard W. Schmude, Jr.)

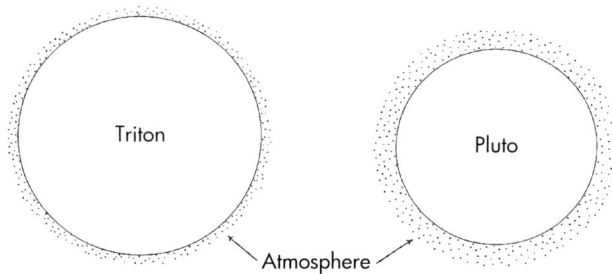

Figure 3.5. A comparison of the atmospheres (shaded areas) of Neptunian moon Triton and Pluto. Since Triton is larger than Pluto, it has a stronger gravitational field and thus its atmosphere lies closer to the surface than Pluto's atmosphere. (Credit: Richard W. Schmude, Jr.)

when temperatures rise. Since the surface temperature undoubtedly falls during the night, part of the atmosphere may also condense onto the surface at this time.

We are not sure about the temperatures of the lowest layers of Pluto's atmosphere. One possibility is that it has a troposphere. This is a cold layer near the surface where the temperature drops gradually with increasing altitude. At some point, perhaps several kilometers above the surface, this trend would stop and reverse. This is the situation on Triton. If this is the case for Pluto, then clouds similar to those on Triton may form near its surface. There is a chance that Pluto has a troposphere during part of its year and, hence, clouds may be seasonal features.

Pluto probably has an ionosphere like Triton. Ultraviolet light from the Sun would ionize molecular and atomic nitrogen creating N_2^+ and N^+. Some of the hydrogen ions (H^+) in the solar wind may also be in Pluto's ionosphere. The ionosphere would also contain free electrons.

Pluto's atmosphere extends to higher altitudes than Triton's atmosphere. On Triton, the atmospheric pressure is 1.0 μbar at an altitude near 40 km whereas the corresponding altitude for Pluto is near 90 km. Figure 3.5 illustrates this difference. There are two reasons for this difference. The first reason is temperature. The higher temperatures of Pluto's atmosphere cause the gases to attain higher altitudes, because, the higher the temperature, the faster the gas molecules are moving and the more likely they are of escaping. A second reason for the higher altitudes of gas on Pluto is that it has a lower mass than Triton. As a result, Pluto's gravity is weaker and once again, it is easier for gases to escape from Pluto than from Triton.

At the present time, material is escaping from Pluto faster than from Triton. One group estimates that 8 billion kilograms of nitrogen escapes from Pluto each year based on Pluto's temperatures in the late 20th century. There is a chance that lots of material has escaped from it over the age of the Solar System, leading to a higher percentage of rock on it. One must remember, though, that Pluto is usually much colder than what it was in the late 20th century when it was near perihelion and, hence, escape rates would be usually much lower than the rate just quoted.

Interior

The average density of Pluto is about half-way between that of water ice and rock; therefore, that body probably contains large amounts of both materials. Other icy materials like carbon monoxide, carbon dioxide and ammonia are probably also

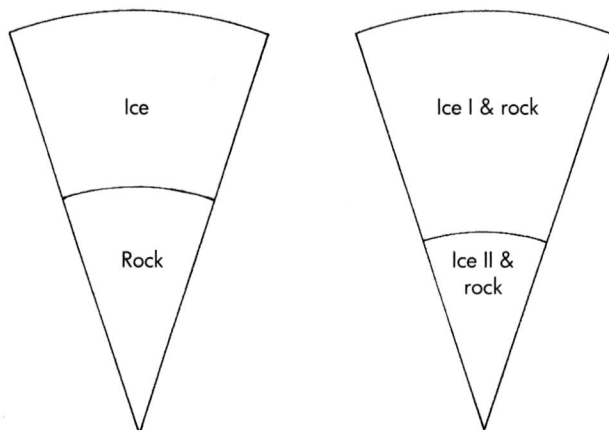

Figure 3.6. Two possible compositions for the interior of Pluto. If Pluto's interior was hot at one time, the water would have separated from the more dense rock portion and the interior would be like the diagram on the left with distinct rock and ice layers. If Pluto's interior was always cold, the water and rock would be mixed together as shown at the right. Due to the higher pressure in the deep interior, Ice-II will become more stable than Ice-I as shown on the right. (Credit: Richard W. Schmude, Jr.)

mixed in with water ice. The rock layer probably contains silicates, oxides, sulfides, metals, salts and some chemically bonded water. The general consensus is that the temperatures of Pluto's interior rose to the point where volatile materials like water and carbon dioxide separated from the rocky core to produce an icy outer layer. See Figure 3.6.

Three possible sources of internal heat for Pluto are: radioactive decay, a giant impact and heat left over from formation. If temperatures have always been cold inside of Pluto, the rock and ice would be mixed as shown on the right side of Figure 3.6. Since Ice-II is more stable at high pressures, it will be present in the interior, whereas Ice-I will form near the surface. Pluto may contain large quantities of carbon-based material similar to what is in carbonaceous chondrite meteorites. If this is the case, that material would lie above the rocky core. A thin outer layer of nitrogen ice lies on Pluto's surface.

One study suggests that Pluto's core may reach a temperature of 1,000 K and its pressure may reach 10,000 bar. The amount of heat escaping Pluto is around 0.003 Joule/m^2 s, which is about 1% of the rate for Neptune. There is a good chance that this heat is released in just a few small areas on Pluto, which will cause hot spots.

Magnetic Environment

Due to Pluto's small size and slow rotation rate, it probably does not have a sizable magnetic dynamo. It, however, may have a small residual magnetic field. Even our Moon has a residual magnetic field. If Pluto's surface magnetic field reaches 0.0002 Gauss (or 0.0007 times that of Earth's field) it would be enough to deflect the solar wind. In this case, Pluto's magnetosphere would look like that in Figure 3.7. If Pluto lacks a magnetic field, the solar wind would interact with its atmosphere.

Another scenario is that Charon has a residual magnetic field whereas Pluto lacks one. In that case, Charon would have the magnetosphere instead of Pluto.

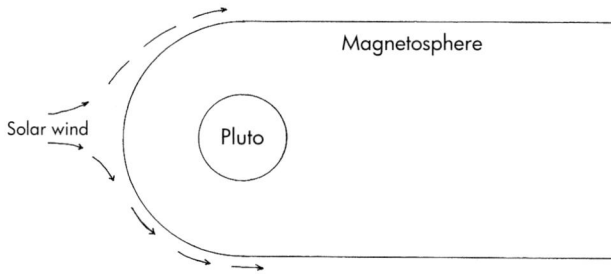

Figure 3.7. A possible layout of Pluto's magnetosphere. A residual magnetic field of just 0.0002 Gauss is sufficient to deflect the solar wind around Pluto, as shown. (Credit: Richard W. Schmude, Jr.)

Surface

How did astronomers determine Pluto's diameter? Unlike the eight major planets, Pluto's disc can not be resolved except by the largest telescopes and even these instruments do not show a sharp round image for Pluto. Astronomers therefore had to resort to other ways of measuring its size.

During the mid- to late-1980s, Pluto and its moon, Charon, moved in front of each other. Astronomers measured the exact times when these events occurred and from this data, they were able to determine approximate diameters for both objects. Three sources of uncertainty that people dealt with when using this method were: albedo features on Pluto and Charon, limb darkening of both objects and the uncertain effect of Pluto's atmosphere on the results.

A second and even more precise method of measuring Pluto's size and shape is stellar occultation measurements. Since we are not sure of Pluto's exact position, we can not determine Pluto's size from occultation data collected at just one site. If, however, data are collected at several sites, we can use the distances from the different sites along with the occultation data to evaluate Pluto's size and shape.

Astronomers at several different sites successfully measured occultations in 1988, 2002, 2006 and 2007. One such event, in June 2006, was measured by several people including at least two amateurs; this is discussed further in Chapter 6.

Pluto and Charon are always within one arc-second of each other and, as a result, brightness measurements generally refer to the combined light from these two objects. The combined light from Pluto's two small moons – Nix and Hydra – is less than 0.1% of the light coming from Pluto; hence, it is negligible. See Figure 3.8. The photometric constants of Pluto + Charon are listed in Table 3.4. Pluto gives off ~84% of the total light and Charon gives off the other 16% from the Pluto-Charon system. Therefore, any change on Pluto will affect the Pluto-Charon brightness.

The amount of light reflected by the Pluto-Charon system has fallen at a rate of about 3% per decade between the 1930s and 2000. This change may be due to the sublimation of nitrogen ice that lies above darker areas. Essentially, as the surface temperature rises the top layer of nitrogen ice sublimes exposing an underlying dark layer. See Figure 3.9. Sublimation of nitrogen would also explain the rise in atmospheric pressure from 1988 to 2002. The current episode of nitrogen sublimation may be due to the fact that Pluto's seasons have changed.

Table 3.4. Photometric Constants for the Pluto + Charon system

Characteristic	Value
V(1,0)	−0.69 [a]
Vo	15.22 [a]
B–V	0.85
V–R	0.46
Average geometric albedo (V filter)	0.49
Solar phase angle coefficient (B filter)	0.037 magnitude/degree

[a] Values for 1999 data which were taken from Buratti et al (2003).

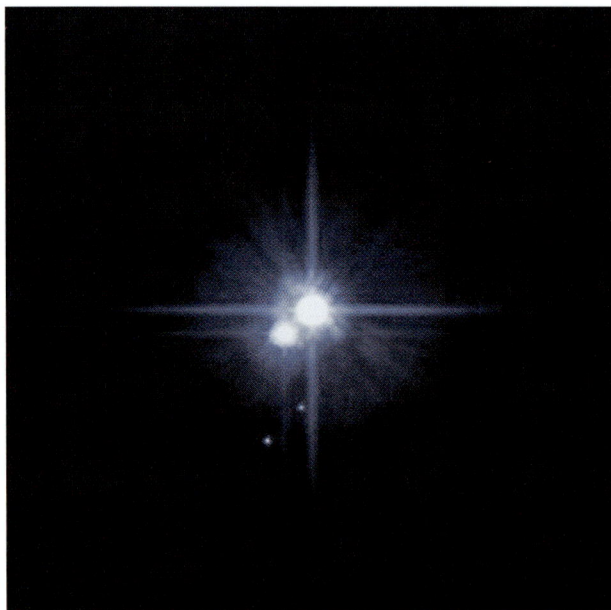

Figure 3.8. A Hubble Space Telescope image of Pluto, Charon, Nix and Hydra. Pluto is the brightest object followed by Charon to its lower left. Nix is the small bright dot that is almost directly below Pluto and Hydra is the second bright dot that is to the lower left of Nix. (Credit: NASA, ESA, H. Weaver, A. Stern and the HST Pluto Companion Search Team.)

A second possible reason for Pluto's falling albedo between 1930 and 2000 is its changing orientation as seen from Earth. In 1930, its northern hemisphere faced us, but by 2000, the southern hemisphere was tipped towards Earth. If Pluto's northern hemisphere is brighter than its southern hemisphere that would explain the falling albedo.

Data taken by Doug West between 2001 and 2004 are consistent with a Pluto + Charon albedo of 0.61 in a filter transformed to the Johnson V system. This is higher than in the 1990 s.

In addition to albedo, Pluto's light curve has also changed over the last 50 years. In the mid-1950 s, its darkest longitude was 230° and it was 0.1 magnitudes

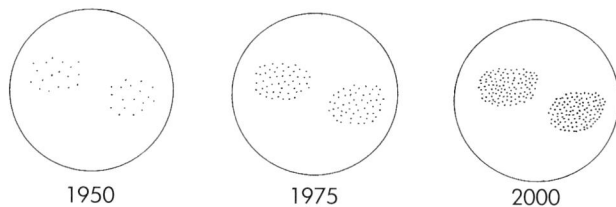

1950   1975   2000

Figure 3.9. One explanation for the falling albedos of Pluto and Charon is the sublimation of nitrogen ice. Essentially, as the overlying nitrogen ice sublimed between 1950 and 2000, darker layers became visible. This may have led to Pluto's diminishing geometric albedo between 1954 and 1999 and to the increasing amplitude of its light curve in visible light between the 1950s and the 1990s. (Credit: Richard W. Schmude, Jr.)

dimmer than the brightest longitude at 100°. During the mid-1960 s, this difference had increased to almost 0.2 magnitudes, and by the 1990 s, this difference had increased to 0.3 magnitudes. The amplitude of Pluto's light curve had more than doubled between 1954 and 1999. When a dark patch on Pluto is visible, brightness drops, but when a hemisphere with almost no dark areas faces Earth, Pluto is brighter. The changing light curve is probably due to the sublimation of bright surface ice; darker areas are then exposed, and this leads to more extreme light curves for it. As temperatures fall, the nitrogen atmosphere will freeze on to the surface and cover up the dark areas. This, in turn, will cause the darkest areas to become brighter and, as a result, the difference between the darkest and brightest longitudes will fall. See Figure 3.9. There is also a chance that the changing orientation of Pluto as seen from Earth is changing the light curve.

Do Pluto and Charon reflect more light as their solar phase angles drop? During the 1980s, this was a difficult question to answer because astronomers could not separate the light coming from Pluto and Charon very well and also because the combined light from these two objects changes as a result of Pluto's rotation. In three studies, based on data from the 1980 s, astronomers used a filter transformed to the Johnson B system and found that Pluto + Charon dim at a rate of 0.03–0.04 magnitudes per degree of phase angle. What this means is that the combined light from these two objects were 0.03–0.04 magnitudes dimmer at a solar phase angle of 1.7° than at 0.7°. In a more recent study, astronomers used Hubble Space Telescope images to separate the light reflected by Pluto and Charon. This group reported that Pluto dimmed by 0.0294 ± 0.011 magnitudes (V filter) for each one degree increase in its solar phase angle, while Charon dimmed at a rate of 0.0866 ± 00078 magnitudes (V filter) for each one degree increase in its solar phase angle. These results suggest that, Charon has a more porous surface than Pluto.

It is difficult to comment on the opposition surges of Pluto and Charon because we only have brightness data covering solar phase angles from ~0.5° to ~2°. One group reports that Uranus' moon, Titania, dims at a rate of 0.102 magnitudes per degree at very low solar phase angles in a filter transformed to the Johnson V system. This rate is just over three times that for Pluto and, hence Pluto's opposition surge is probably one-third that of Titania or ~0.1 magnitudes.

One must remember that Pluto's photometric constants may change with its season. In the year 2100, for example, much of Pluto's surface may be covered by a thin layer of ice from the atmosphere. This would probably cause the opposition surge and solar phase angle coefficient to drop.

Unlike Triton, Pluto has large areas with V filter albedos below 0.5. The darker areas are a little redder than the brighter ones. This is consistent with darker areas having a higher concentration of hydrocarbons, which are both darker and redder than nitrogen ice. Parts of Pluto's surface may have been darkened by charged particles colliding with the surface ices creating darker compounds. Both galactic cosmic rays and solar wind particles can darken Pluto's surface ices.

What is the temperature of Pluto's surface and does it change? We are not sure of the exact temperature of Pluto's surface, but we have an idea of what is. We know how much energy the Sun gives off and by using this along with Pluto's distance and albedo, we know that it is around 37 K. At this temperature, nitrogen has a vapor pressure of a few microbars, which is probably near Pluto's surface pressure. Some of the darker areas may rise to 50 K. These are believed to be free of nitrogen ice. Methane may sublime in these areas. The vapor pressure of this compound at 50 K is 2 µbar and, hence, Pluto's atmosphere may contain a significant amount of methane near perihelion. Areas even warmer than 50 K may be present on Pluto. These areas could be heated by the escape of internal heat or could have low albedos and be heated by sunlight.

Many astronomers believe that the ices on Pluto's surface are in equilibrium with the atmosphere. If this is the case then changes in the surface temperature will lead to changes in Pluto's atmosphere. Essentially higher temperatures will lead to higher atmospheric pressures at the surface. In fact, between 1988 and 2002, the average surface temperature may have risen by about one Kelvin causing more nitrogen to sublime into Pluto's atmosphere. This would lead to a higher surface pressure, which was observed.

Figure 3.10 shows a possible cross-section of Pluto's surface. Water ice lies underground and makes up the "bedrock" layer. Ammonia and other impurities are probably in the water ice layer. A layer of nitrogen ice mixed with carbon monoxide and methane lies above the water ice. This layer is probably darker than fresh nitrogen ice since it contains impurities. Small spots of fresh nitrogen ice may also lie on the surface. Large quantities of tholins – darker hydrocarbon molecules – probably cover large regions. This material probably comes from both the settling of atmospheric hazes and from solar radiation breaking up the methane molecules on the surface. One group reports that frozen ethane is also present. Argon ice may also lie on Pluto. Argon makes up about 1% of our

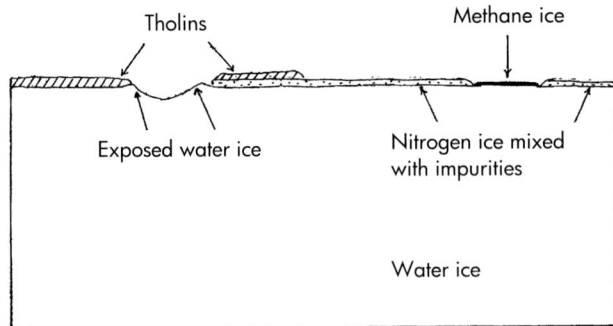

Figure 3.10. A possible cross-section of Pluto's surface. Nitrogen ice mixed with impurities probably covers much of Pluto's surface. Areas of exposed methane ice, water ice and tholins may also be visible from Earth. A layer of water ice may lie below Pluto's surface. (Credit: Richard W. Schmude, Jr.)

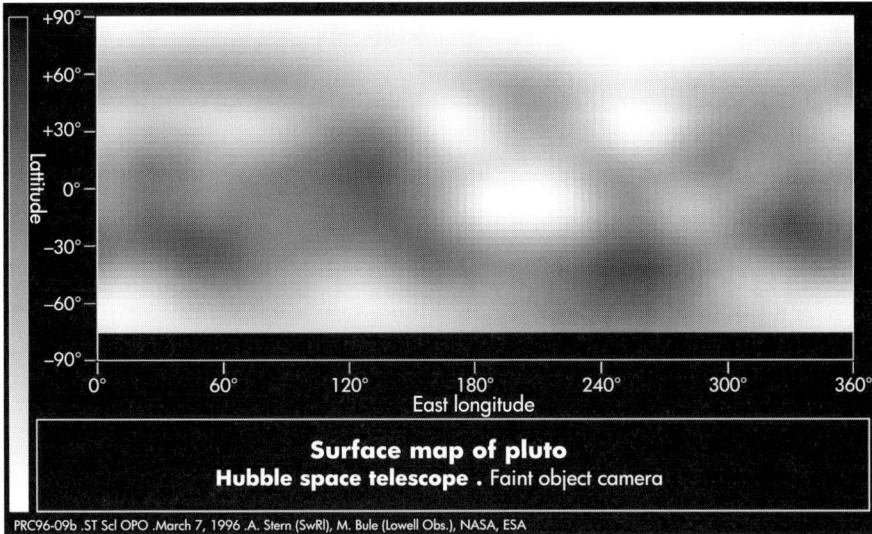

Figure 3.11. A map of Pluto constructed from Hubble Space Telescope images taken in June and July 1994. The brightness variations may be due to topographic features and the distribution of different ices on Pluto. Positive latitudes refer to the northern hemisphere. (Credit: Alan Stern, Marc Buie, NASA and ESA.)

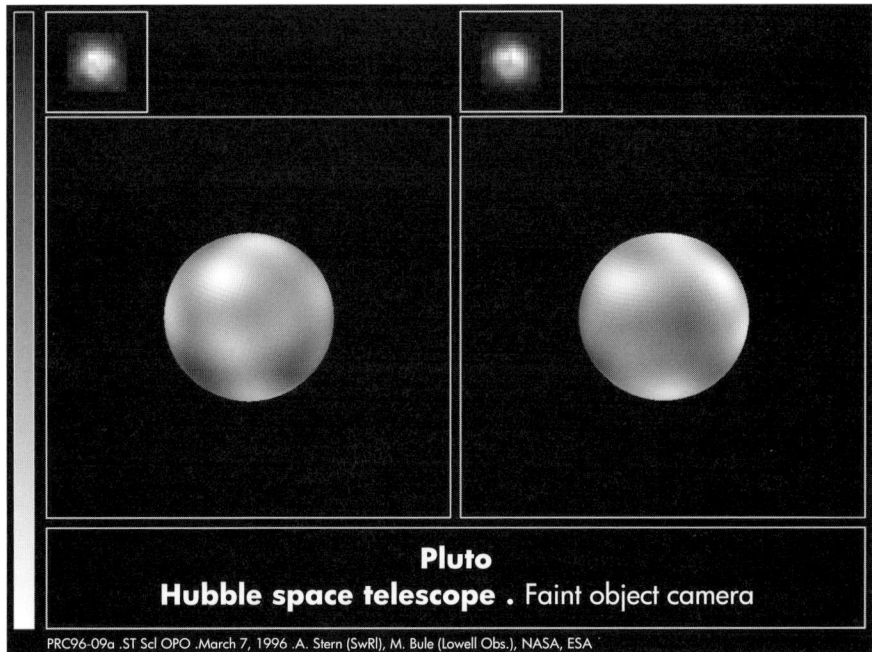

Figure 3.12. Two images of Pluto taken with the European Space Agency's (ESA) faint object camera on the Hubble Space Telescope. North is near the top. (Credit: Alan Stern, Marc Buie, NASA and ESA.)

atmosphere. Much of the surface is probably quite smooth; however, there are probably craters on Pluto's surface along with some debris.

The nitrogen ice will remain in the beta phase until the surface temperature falls below 35.6 K. This may not occur until after 2050. When temperatures drop below 35.6 K, the denser alpha phase will form, and this may cause cracks in the surface ice.

Figure 3.11 shows a map of Pluto. Astronomers constructed this map from Hubble Space Telescope images taken with the faint object camera that was built by the European Space Agency (ESA). Figure 3.12 shows two different sides of Pluto; the faint object camera on the Hubble Space Telescope took both images. It will be interesting to see if Pluto's dark spots grow from 1996 to 2015.

Satellites

As of late 2007, Pluto had three known moons: Charon, Nix and Hydra. Characteristics of all three are summarized in Table 3.5. Figure 3.13 shows the relative sizes of Pluto and Charon and the orbits of Pluto's moons.

Charon

Charon is Pluto's largest moon. It is the 12th largest satellite in our Solar System. If Charon has an atmosphere, its surface pressure would be less than 0.2 μbar. Since Charon has only one-eighth the mass of Pluto, its gravity is weaker and, hence, gases are able to escape more easily. Therefore even if an atmosphere developed on this moon, it would escape.

Table 3.5. Characteristics of Pluto's three moons

Characteristic	Charon [a]	Nix [b]	Hydra [b]
Radius (km)	605	40 ?	40 ?
Period of revolution (days)	6.387	24.856	38.207
Semi-major axis (km)	19,571	48,675	64,780
Vo	17.26	24.55	24.39
V(1,0)	1.35	8.64	8.48
Geometric albedo (V filter)	0.35	0.1 ?	0.1 ?
B–V	0.71	0.91	0.64
Surface composition	H_2O and NH_3 hydrate	?	?
Color	Gray	Slight yellow	Gray
Mass (10^{18} kg)	1,520	≤1 [c]	≤1 [c]
Density (g/cm^3)	1.7	1.7 ?	1.7 ?
Surface Temperature	~45 K	~50 K	~50 K

[a] The radius, mass and density are estimated from the values in Person et al (2006). The Vo, value is from Stern and Tholen (1997) p. 203; the B–V value is from Stern and Tholen (1997), p. 288 and the temperature is from Cook et al (2007).

[b] The Vo, B–V, semi-major axis and period of revolution values are from Buie et al (2006).

[c] Tholen et al (2007) report an upper limit of 0.07 kg^3 s^2 for the GM values of Nix and Hydra, where G is the gravitational constant and M is the satellite mass.

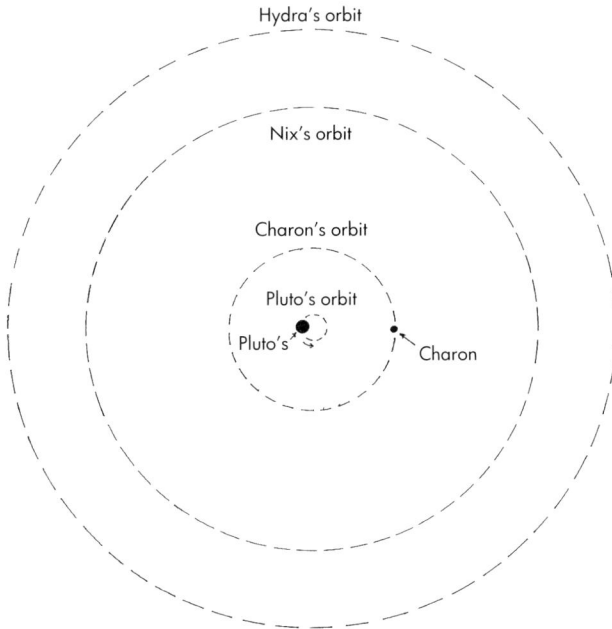

Figure 3.13. Relative sizes of the orbits of Charon, Nix and Hydra, along with that of Pluto and Charon. Pluto and Charon are the black circles. Pluto moves in the small orbit around the center of mass. This is what causes it to wobble. (Credit: Richard W. Schmude, Jr.)

Both theoretical studies and Charon's light curve are consistent with it undergoing synchronous rotation; therefore, the same side of Charon probably faces Pluto just like the same side of our Moon always faces Earth. The sub-Pluto point on Charon is defined as the $0°$ longitude point. As Charon rotates, its brightness changes by a few percent; it is dimmest on the side opposite of Pluto.

Water ice, in a crystalline state, is on Charon's surface. Water covers much of the surface. One group reports that it might be a little more abundant on the leading hemisphere than on the following hemisphere. The crystalline state and distribution of water ice on Charon is similar to that on Uranus' three large moons Titania, Umbriel and Ariel. One group reports that ammonia hydrate is also present on Charon. This compound is probably mixed with the water ice. Methane, nitrogen and carbon monoxide may be present but are less abundant than on Pluto. A dark material with a neutral color is mixed with the water ice; this material causes Charon's albedo to be lower than that of pure ice. This material may be similar to the tholins on Pluto. Most of the pure methane and nitrogen on Charon has probably escaped into outer space.

Charon reflects about 38% of the blue light and about 36% of the green and red light falling on it. These values are lower than those for Pluto and, hence, Charon absorbs a higher percentage of sunlight than Pluto. This is one reason why Charon's surface temperature may be higher than Pluto's. Pluto probably reflects more light than Charon because it receives a fresh coat of nitrogen ice every time the temperature drops. This ice comes from the atmosphere. Any ice forming on Charon from its atmosphere would be much thinner than the corresponding layer on Pluto and would have little or no effect on the albedo.

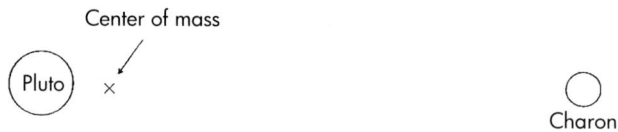

Figure 3.14. The center of mass of Pluto and Charon. Since Pluto has a higher mass than Charon, the center of mass lies closer to Pluto. Note that the center of mass lies above Pluto's surface. (Credit: Richard W. Schmude, Jr.)

How did astronomers measure the masses of Pluto and Charon without space probes? Astronomers first used Kepler's Third Law of Planetary Motion along with the Pluto-Charon distance to compute the combined mass of Pluto and Charon. They then measured Pluto's wobble as Charon revolved and, from this, determined individual masses. Essentially Pluto and Charon revolve around the center of mass of the Pluto-Charon system and, as a result, Pluto wobbles by about 0.2 arc-seconds every 6.387 days. Astronomers used the size of the wobble to pinpoint the location of the center of mass, which is point × in Figure 3.14. They computed the mass of Pluto using equation 3.1:

$$M_P \times D_1 = M_C \times D_2 \qquad (3.1)$$

where M_P and M_C are the masses of Pluto and Charon, D_1 is the distance between Pluto's center and point × and D_2 is the distance between Charon's center and point ×. Pluto's wobble equals twice the value of D_1 as seen from Earth. Once astronomers knew Pluto's mass and the combined mass of the Pluto-Charon system, they were able to compute Charon's mass.

Does Charon have some kind of cryovolcanic activity? Possibly; let me explain. Cryovolcanic activity occurs when material with a low melting point, like water, covers a very cold surface. One group argues that sunlight and cosmic rays should turn Charon's layer of crystalline water ice into amorphous water ice. (As the reader will recall, crystallized ice contains water molecules that are arranged in an organized pattern, whereas amorphous ice contains water molecules that are arranged in a random pattern.) Since crystalline ice is present, something must be replenishing it. One possibility is a liquid water-ammonia mixture that seeps up to the surface. A more definite answer to cryovolcanic activity, however, can not be made until we have high-resolution images of Charon.

How can a cold body like Charon have volcanic activity? Since Charon has a density of 1.7 g/cm³, its interior must possess large amounts of rock. One group points out that Charon's interior is probably heated by radioactive elements. They predict that the core may generate enough heat to melt some of the deep ice layers resulting in a liquid layer near the rocky core. They go on to suggest that some of this liquid may reach the surface and replenish the crystalline ice.

Nix, Hydra, and Beyond

Nix and Hydra both orbit the center of mass of the Pluto-Charon system. See Figures 3.13 and 3.14. As it turns out, the center of mass lies in outer space, and, hence, one should not treat Pluto as a point mass. Essentially, one must use more complicated models than Kepler's Third Law to study the movement of these two moons.

One group reports that both moons do not change in brightness as they revolve around the Pluto-Charon center of mass. This is consistent with the moons having nearly circular shapes, which imply larger diameters and lower albedos. A spherical moon will have nearly the same area as it revolves, whereas an irregular moon will have different areas projected towards Earth which will cause brightness changes. As a result, I have assumed a low albedo of 0.1 for both moons, which will yield larger sizes and masses. Physical characteristics of each moon are listed in Table 3.5.

Both Nix and Hydra move in nearly circular orbits, with eccentricities below 0.01. Their orbits lie in nearly the same plane as Charon's orbit. We are not sure if they undergo synchronous rotation.

Does Pluto have additional moons? There is a chance that small moons below our detection limit are present. Pluto's Hill Sphere extends to beyond 5 million kilometers. (The Hill Sphere is described at the end of Chapter 1.) Moons farther than Hydra may be present. If Pluto has additional satellites beyond Nix and Hydra they would be dimmer than 25th magnitude.

We know from occultation data that Pluto does not have thick rings like those around Saturn and Uranus; however, it may have thin rings like those orbiting Jupiter and Neptune.

Chapter 4

Observing Uranus and Neptune with Binoculars and Small Telescopes

This chapter describes how one can estimate the brightness and color of Uranus and Neptune with either binoculars or a telescope having a diameter of under 0.1 meters (4 inches). (Larger instruments are needed for Pluto. This and other matters are described in the next chapter.) This chapter starts with a description of the human eye, followed by discussions on binoculars, telescopes, how to find Uranus and Neptune in the sky, using the visual brightness method, and making visual magnitude estimates and color estimates. This chapter contains also a list of comparison stars for both visual magnitude estimates and photoelectric photometry magnitude studies of these planets. Photoelectric photometry is discussed in the next chapter.

The Human Eye

The human eye is a remarkable organ. A diagram of the eye is shown in Figure 4.1. I will describe how the lens, pupil, retina, and fovea centralis work. These descriptions should enable the observer to make better visual observations and to know the limitations of their own eyes.

The eye contains a flexible lens. When one focuses on a close object, the lens changes shape to focus the light on the fovea centralis; this is different from a telescope, where focus is achieved by changing the distance between optical surfaces. While the human eye lens has imperfections, the bad effects of these imperfections are reduced when one uses both eyes to observe.

The iris is in front of the lens and determines the size of the pupil. The larger the size of the pupil, the more light that enters the eye. When there is little light, the pupils enlarge to let in more light, but when there is a lot of light, the pupils contract. The maximum pupil size varies from person to person and with age. The average dark-adapted pupil size for different ages is about 7.0 mm (13 years old), 6.0 mm (35 years old), 5.0 mm (57 years old) and 4.0 mm (80 years old). As a general rule, the older one is, the smaller will be his or her dark-adapted pupil. In some cases, however, a 60-year-old may have larger dark-adapted pupils than a 20-year-old. When the light intensity suddenly drops, the rods in the retina are unable to detect light because they lack the important chemical, rhodopsin, and, as a result, they do not function properly. After a person has been in the dark for several minutes, the rods begin to adjust to the lower light levels. Once the rods have adjusted fully, the eyes become dark-adapted. At least 20 to 30 minutes is

R.W. Schmude, Jr., *Uranus, Neptune, and Pluto and How to Observe Them*,
DOI: 10.1007/978-0-387-76602-7_4, © Springer Science+Business Media, LLC 2008

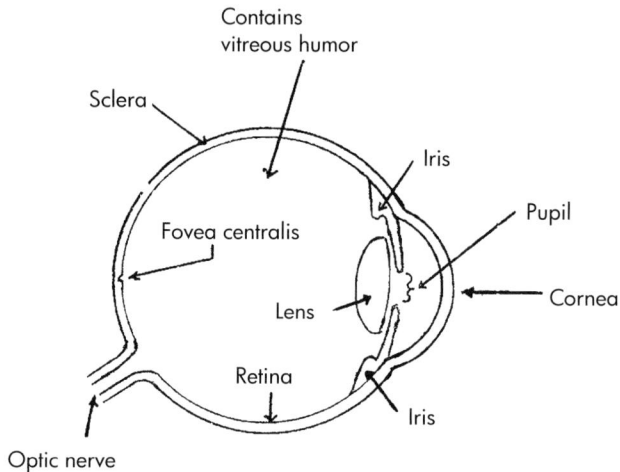

Figure 4.1 Diagram of the human eye. (Credit: Richard W. Schmude, Jr.)

required before one gets dark-adapted fully, but some people may need 60 minutes, especially if they were exposed to bright light earlier in the day.

The retina in each eye contains ~120 million rods and ~6 million cones. The rods and cones are the detectors; they absorb incoming light and send signals to the brain where an image is constructed. The cones have diameters of between 1.5 and 8 μm, depending on their location on the retina. There are three types of color pigments in the cones, which are called S (small wavelength), M (medium wavelength) and L (large wavelength), with peak sensitivities at wavelengths near 430, 530 and 560 nm (nanometers). The S cones are sensitive to blue light, the M cones are sensitive to blue and green light and the L cones are sensitive to green, yellow, orange and red light. When light enters the eye and strikes the cones, signals from the three types of cones are sent to the brain and a color image is constructed. An electronic camera operates in much the same way – different colors are constructed from a combination of red, green and blue light. Cones need lots of light to function. Consequently, people do not see colors in dimly lit areas. Rods, on the other hand, are our receptors of faint light, and have a thousand times the maximal sensitivity of cones in the dark-adapted eye. However, rods do not resolve fine detail such as the print on this page. Hence, it is difficult to read in dim light. Rods do not distinguish color, but rather enable one to see in shades of gray in dimly lit areas.

The light intensity affects the perceived color of an object. The full moon, for example, appears to have a white color even though its true color is yellow-white. This is because it reflects all colors of light, and, when one looks at it, his or her cones are bombarded by large amounts of all colors of light. The end result is that the cones reach maximum stimulation in many colors, and the perceived color of the full moon is white instead of yellow-white. When one looks at Saturn with the unaided-eye the cones do not attain maximum stimulation, and, as a result, the yellow-white color is perceived. As it turns out, Saturn and the full moon have nearly the same color.

Not all things that we see are in sharp focus. When one reads, he/she must bring each word into focus; or, in other words, as one reads, his/her eyes scan the page. The reason for this is that each word must be focused on the fovea centralis, which has an area of less than one square millimeter. The fovea centralis contains only cones, and when one focuses on this area, high-resolution-color-vision results. Our view of objects outside of our focus area lacks some color detail and resolution since there is a lower density of cones outside of the fovea centralis. Applying this to observing astronomical objects, our Moon is large enough that only a small part of its image will fall within the fovea centralis when seen through 10×50 binoculars. Hence, we have to scan it to see in the best focus. On the other hand, when we look at Uranus through a telescope, even at a very high magnification of 500X, the entire image of the planet falls into the fovea centralis and appears to be in sharp focus. However, the peripheral areas of the view, where the planet's moons are found, lie outside of the fovea when we direct our gaze at the planet.

The best way to view a faint object is to use averted vision; this is, where one does not focus directly on the object, but instead, on an object a short distance away and lets the light from the faint object fall on an area of the retina with more rods. As stated earlier, rods, and not cones, are better suited to detect faint light levels.

Light Sensitivity

The human eye is able to detect light with wavelengths between about 410 and 710 nanometers (nm). This range, however, varies from person to person. When there is lots of light, our eyes are most sensitive to green light with a wavelength of 550 nm. At low light levels, when the rods are the primary detectors, our eyes are most sensitive to blue-green light with a wavelength near 510 nm. At intermediate light levels, such as when one views Uranus through a telescope, the eyes may be most sensitive to light with a wavelength of 530 nm. Estimating the colors of Uranus and Neptune must be done always under nearly the same lighting since the eye's peak sensitivity changes under different lighting conditions. Color is discussed later.

The faintest star that the eye can detect depends on the size of the dark-adapted pupil, the background lighting and the color of the star. For most people, a dark-adapted eye is able to see objects that are brighter than about magnitude 6.0.

One must always be careful about the eye's differing color sensitivity. A red magnitude 5.0 star will almost always appear fainter than a yellow magnitude 5.0 star because the eye is more sensitive to yellow than to red light. In fact, this was a serious problem with 19th century star catalogs. Essentially, red stars were recorded as being fainter than equally bright blue stars. Differences in color between a comparison star and a target like Uranus can be a source of systematic error for visual magnitude estimates.

One must also avoid staring at a star because in doing so, he or she will get a different response than merely glancing at it. This is because when one stares at a star, one is focusing the light on the fovea centralis, which has a different color sensitivity than the rest of the retina. When one glances at a star one is focusing the light on the retina and not the fovea centralis.

Our eyes have different color sensitivities than commonly used filters, such as those transformed to the Johnson B, V, R and I system. Therefore, stars with a V filter magnitude of 4.0 as measured with a photoelectric photometer may not appear as a magnitude 4.0 star to our eyes. In fact, a red star can appear 0.2 or more magnitudes fainter to the eye than the listed V filter magnitude. Several people have examined this problem. I would like to discuss a recent experiment carried out by Richard Stanton for the American Association of Variable Star Observers (AAVSO).

Stanton carried out an experiment to determine how the "average observer" records star brightness and how these values compare to those measured with a V filter and a photoelectric photometer. A total of 63 people sent in over 650 individual observations of magnitude 7 to magnitude 15 stars using a list of comparison stars of known color and brightness. The observers had a range of observing experience ranging from beginner to experienced observer. Stanton made several important findings from this study, three of which were:

1. When visual magnitudes of the same star made by different people are combined, a standard deviation of around 0.2 magnitudes can be expected;
2. There is no need to report visual magnitudes to increments less than 0.1 magnitudes; and
3. One can compute a visual magnitude, m_v, from a Johnson V filter magnitude and the color index, (B–V) through:

$$m_v = V + b \times (B - V) \tag{4.1}$$

where b is a constant to be determined for each observer, V is the Johnson V filter magnitude and B is the Johnson B filter magnitude. Stanton computed b values for each of his observers with the result that the average b value for all 63 observers was 0.21 ± 0.01.

Binoculars

Binoculars are not just for beginners! Binoculars can be used to study variable stars, deep sky objects, Uranus, and Neptune. The three advantages of binoculars are that they are portable, they allow the observer to use both of his or her eyes, and it is easier to find things with binoculars than with a telescope. You should almost always use binoculars to find Uranus and Neptune before using the telescope to locate them. One additional advantage of binoculars is that they require almost no set-up or tear-down time. One simply points and looks.

There are three important numbers for binoculars, namely, the magnification, the diameter of the objective lenses and the field-of-view (FOV). The first two numbers are always stated while the FOV may or may not be stated. As an example, 7×35 binoculars have a magnification of 7 power and have objective lenses with a diameter of 35 mm. The FOV describes the angular size of the view as seen through the binoculars. Normal eyes have an FOV of about 170° whereas a typical pair of binoculars has an FOV of around 3° to 8°.

Modern binoculars have two basic designs: roof prism and porro prism. See Figures 4.2 and 4.3, respectively. In the roof prism design, the eyepieces are directly in front of the objective lenses. This is not the case in the porro prism design. In both binocular designs, the light comes in and bounces off the prisms several times before entering the eyepieces. Roof prism binoculars are usually light and more compact than porro prism binoculars, but are often more expensive. Both designs use prisms that cause the image to be the same as what we see with our eyes. Telescopes usually flip the image upside down and/or invert the image

Figure 4.2 A pair of roof prism binoculars. Note that the eyepieces at top are directly in front of the objective lenses. (Credit: Richard W. Schmude, Jr.)

Figure 4.3 A pair of porro prism binoculars. Note that the eyepieces at top are not directly in front of the objective lenses. (Credit: Richard W. Schmude, Jr.)

from right to left. One should also remember that a finderscope may show a different orientation of the sky than binoculars.

Most modern binoculars have coatings on their lenses. This is usually a thin layer of magnesium fluoride. The purpose of the coatings is to increase the amount of light traveling through the optical surfaces by reducing the amount of reflected light. Light strikes the uncoated glass in Figure 4.4 and some of it is reflected while the remainder passes through or is absorbed by the glass. When light strikes the coated glass in Figure 4.5, more of it passes through the glass because less of it is lost due to reflection. The best type of coating is fully multicoated, which means that each lens-to-air surface is coated with several thin layers of the coating. This causes less than 1% of the light to be lost to reflection. One must protect the lens coatings from fingerprints and the elements. Lens caps should be used for this purpose.

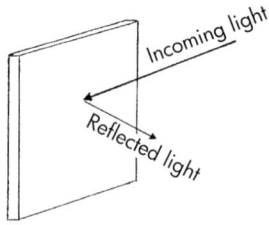

Figure 4.4 Uncoated glass reflects some light. Lenses without coatings transmit less light than coated lenses and, as a result, celestial objects appear dimmer through uncoated lenses. (Credit: Richard W. Schmude, Jr.)

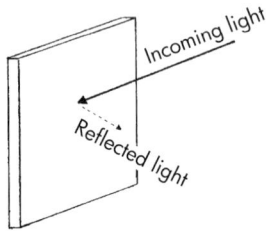

Figure 4.5 Coated glass reflects less light than uncoated glass. Lenses that have coatings transmit more light and cause objects to be brighter. (Credit: Richard W. Schmude, Jr.)

There are several things that one should consider before purchasing a pair of binoculars.

One consideration is weight. A light pair is easier to hold than a heavy pair and will result is less muscle strain. I own two pairs of giant binoculars; one pair weighs three pounds and the second pair weighs five pounds. I use the lighter pair more often.

A second consideration is being able to hold the binoculars steady. The higher the magnification, the more sensitive the binoculars will be to natural movement. I have been able to make thousands of magnitude estimates with my three-pound pair of binoculars by simply placing my elbows on my car when holding the binoculars. The proper way of holding binoculars is shown in Figure 4.6. This stable position has allowed me to keep my binoculars steady. It is important to

Figure 4.6 The correct way of holding a pair of heavy binoculars. Note that the observer is holding the binoculars near their center of mass and his elbows are on a solid surface. (Credit: Timothy Abbott and Richard W. Schmude, Jr.)

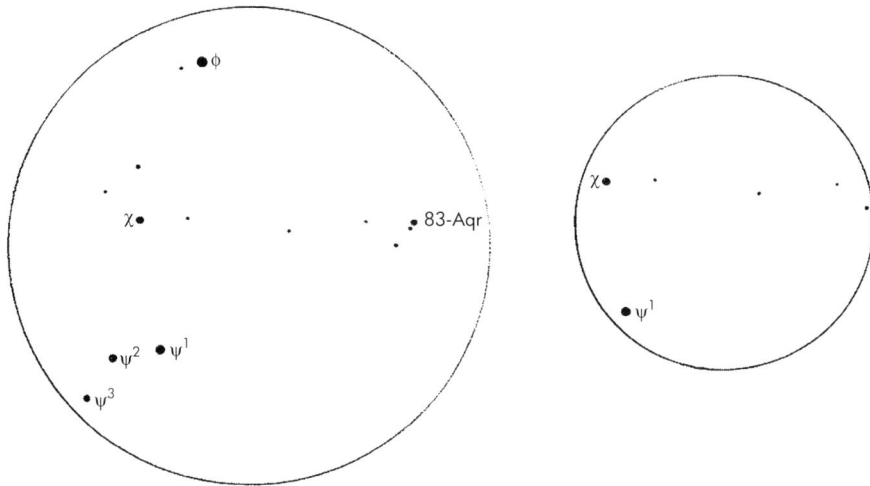

Figure 4.7 The large field-of-view (FOV) on the left shows a nice pattern of stars whereas some of the stars in that pattern are not visible in the smaller FOV on the right. (Credit: Richard W. Schmude, Jr.)

hold the binoculars near their center of mass as in Figure 4.6. Avoid holding the binoculars too close to the objective lenses or too close to the eyepieces.

A third consideration is the field-of-view (FOV). See Figure 4.7. A large FOV shows a larger part of the sky than a smaller one. Beginners need a larger FOV, since it is better for locating objects, whereas the more seasoned observer will not need the large FOV. As a general rule, the higher the magnification, the smaller will be the FOV.

A fourth consideration is the diameter of the dark-adapted pupil. As mentioned earlier, the pupil sizes of most of us decline with age. If the pupil size is small, it may be too small to allow all of the light transmitted through the binoculars to fit through it. The diameter of the beam of light from the binoculars is easily calculated. It is equal to the diameter of the objective lens divided by the magnification. For example, 7×35 binoculars have a beam diameter of 35 mm divided by 7, or 5 mm. This quotient is often called the "exit pupil" of the binoculars, so 7×35 binoculars are said to have an exit pupil of 5 mm. For a 45-year-old observer, an exit pupil of five or six millimeters is fine.

People who wear eyeglasses may want to check to see that the binoculars focus both with and without eyeglasses. In many cases, eyeglasses get scratched and this can degrade the view. There have been many times when the best view is obtained without glasses. Eyeglasses also have extra optical surfaces where light is lost due to reflection and absorption.

One final consideration is the visibility factor for binoculars. The visibility factor is the overall performance number; the higher the number, the higher the performance. It is computed by multiplying the diameter of the objective lenses by the magnification; for example, the visibility factor for 7×35 binoculars is 7 times 35, which is 245. The higher the visibility factor, the better will be the performance if all other factors are equal.

There are several different sizes of image-stabilized binoculars. These binoculars eliminate the shake that occurs from the natural movement of one's arms and thus eliminate the need for a binocular mount. I was very impressed with my view through a pair of image-stabilized binoculars. The visibility factor for these binoculars is high because they have a high magnification.

There are several quality checks that one should do before purchasing a pair of binoculars. One test is to look through the binoculars on a clear night and focus them on a star. The binoculars should yield a sharp focus without any play in the joints. Look at the focused star image: is it a pinpoint (like it should be) or does it show tiny rays sticking out? If the binoculars do not focus a star to a pinpoint of light, they would have astigmatism, which is a problem. Each barrel should point to the same direction; or, in other words, the binoculars should focus into a single image. If they give a double image even for a few seconds, they should be rejected. One can check for dirt by turning the binoculars over and looking down through the objective lenses. Dirt on the inside would be a problem. The binoculars should come with lens caps and a case. These will serve as protection from scratches and small bumps. Finally, the price should be considered. A person new to astronomy may want to start off slowly with perhaps a $100 pair of binoculars rather than a $1,000 pair.

What size of binoculars should one purchase to study Uranus and Neptune? The answer depends on several factors, including eyesight and sky darkness. For most people, I would recommend a pair of 70 or 80 mm binoculars; however, if one is just interested in Uranus, a pair of 35 or 50 mm binoculars should be fine. There are many who would say that one can use smaller binoculars for Neptune, but I disagree. Since the early 1990s, I have received hundreds of Neptune magnitude estimates, and almost all of them were made with 70 or 80 mm binoculars. I have also made over 50,000 variable star magnitude estimates with 11 × 80 binoculars under fairly dark skies at an altitude of 800 feet above sea level. I rarely saw stars fainter than magnitude 9.5 and, in most cases, the magnitude limit was closer to 9.0 under clear skies. Considering that Neptune is magnitude 7.7 and a fainter comparison star is needed, I recommend the larger binoculars for making routine magnitude estimates of Neptune. People who have access to dark skies at elevations above 1 km or who are blessed with excellent eyesight will be able to study Neptune with smaller binoculars. Because of their excellent visibility factor, one may use a pair of ∼45 mm image stabilized binoculars to estimate Neptune's brightness.

Telescopes, Finders, and Eyepieces

The three functions of a telescope from most to least important are: (1) light gathering power, (2) resolution and (3) magnification. Almost all of the light that enters the objective lens or strikes the main mirror of a telescope enters the pupil of the observer's eye or a detector (electronic device, film, etc.). Without a telescope, the only light entering the eye is what passes through the pupil, which is usually between 0.2 and 0.9 cm across. See Figure 4.8. In most cases, one wants to get as much light as possible to improve the accuracy of the measurement. Furthermore, the more light that one has, the higher the magnification they can use. The light gathering power (LGP) of a telescope is expressed as:

$$LGP = d^2 \div p^2 \qquad (4.2)$$

where d is the telescope diameter in meters (m) and p is the pupil diameter in meters. There are 1000 millimeters (mm) in 1 meter and 39.37 inches in a meter. As an example, a 20 cm (8 inch) telescope has an LGP value of 1600 when compared to a pupil diameter of 0.005 m. Since the telescope gathers more light, it enables the

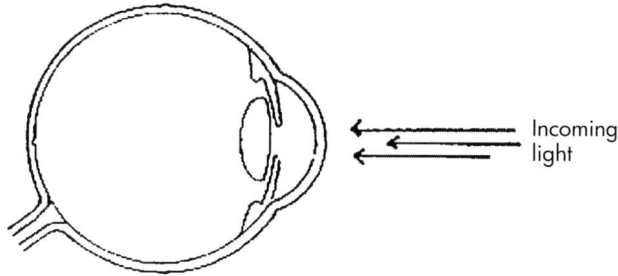

Figure 4.8 The only light that enters our eye is what passes through the pupil as shown here. (Credit: Richard W. Schmude, Jr.)

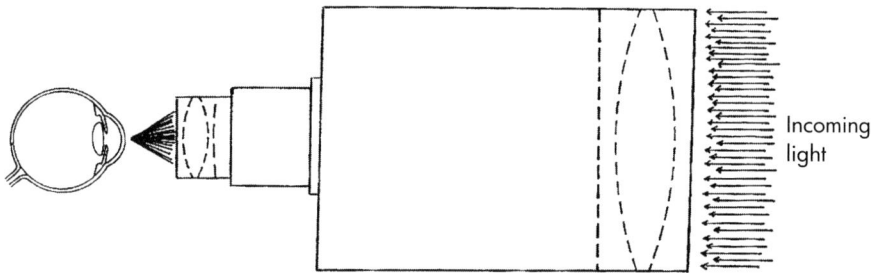

Figure 4.9 All of the light entering the telescope enters the observer's eyes, provided the telescope is properly focused. (Credit: Richard W. Schmude, Jr.)

observer to see fainter objects. See Figure 4.9. A telescope also enables electronic equipment to record a higher signal, which improves the signal-to-noise ratio. Resolution is related to a sharp focus; the lower the resolution value, the sharper will be the focus. The resolution value (R) in arc-seconds is computed as:

$$R = 4.57 \div d \tag{4.3}$$

where d is the telescope diameter in inches. Table 4.1 lists resolution and LGP values for different telescope diameters. As a comparison, the human eye

Table 4.1. Theoretical resolutions and relative light-gathering power of different telescope apertures

Telescope Diameter meters (inches)	Resolution Value–Dawes Limit (arc-seconds)	Light-Gathering Power (LPG) Compared to 0.5 cm Pupils
0.1 (4)	1.1	410
0.15 (6)	0.76	930
0.20 (8)	0.57	1,700
0.25 (10)	0.46	2,600
0.36 (14)	0.33	5,100
0.51 (20)	0.23	10,000
0.61 (24)	0.19	15,000
0.76 (30)	0.15	23,000
1.0 (40)	0.11	41,000

(20-20 vision) has a resolution value of about 60 arc-seconds. There are 3,600 arc-seconds in one degree of arc. The telescope magnification is computed as:

$$\text{Magnification} = \text{telescope focal length} \div \text{eyepiece focal length} \qquad (4.4)$$

As an example, the focal length of my 4 inch refractor is 1.0 meter (which equals 1,000 mm); hence, a 4.0 mm eyepiece will yield a magnification of:

$$\text{Magnification} = 1,000\,\text{mm} \div 4\,\text{mm} = 250\,\text{X}.$$

Remember that the telescope and eyepiece focal lengths must have the same units. The higher the eyepiece focal length, the lower will be the magnification.

There are three common types of telescopes; these are refractors, reflectors (of various types, including Newtonians), and catadioptrics (such as Schmidt-Cassegrains). Table 4.2 summarizes the strengths and weaknesses of each type. Any of the three telescope types are suitable for research of remote planets.

Telescopes almost always have a small FOV, and, as a result, it is difficult to locate objects through them (except in the case of an automated telescope). For this reason, a small telescope with a broad FOV is attached to the main telescope. Its broad FOV enables the observer to find his or her target with it and since it points in the same direction as the main telescope, the target will then appear in the main scope's FOV. In my remote planets studies, the finderscope has been a key accessory. The ideal finderscope should have a magnification of 6 to 12× and have a diameter of at least 50 mm (2 inches). It should have the same orientation as the telescope. The right-angle finderscope in Figure 4.10 gives the same view as a refractor with a star diagonal. A straight finder usually gives the same view that a Newtonian gives. The dew-shield on the finderscope should be at least twice its diameter. If this is not the case, make a dew-shield out of light cardboard and place it in front of the finderscope as is shown in Figure 4.11.

The Telrad finder is also popular. In using it, the observer has a unaided-eye view of the sky, but with a set of concentric circles seemingly projected onto the view so that one can zero in on the target. I have successfully used it on many occasions. When using a Telrad, keep the illuminated portion as dim as possible – this makes the target more obvious and it prolongs battery life. Three limitations

Table 4.2. Strengths and weaknesses of three common types of telescopes

Type	Strengths	Weaknesses
Refractor	Closed tube, low maintenance, convenient location of eyepiece	Lens absorbs ultraviolet light; diameters larger than 7 inches are not portable, expensive per inch of aperture
Newtonian	Low-cost per inch of aperture; can be used for ultraviolet photometry	Periodic maintenance required, difficult to collimate for low f-numbers (focal ratios), eyepiece can be in an inconvenient location
SCT or Maksutov	Low-cost per inch of aperture, closed tube, long focal length, portable, eyepiece is in a convenient location	May require more time to reach ambient temperature than Newtonians

Figure 4.10 A right-angle finder scope. (Credit: Richard W. Schmude, Jr.)

Figure 4.11 A homemade dew-shield on a finder scope. (Credit: Richard W. Schmude, Jr.)

of this type of finder are: (1) one cannot see stars fainter than what can be seen with the unaided eye, (2) it is more susceptible to dew than a traditional finder and (3) it will probably not have the same orientation as the telescope.

The eyepiece is the small lens at the end of the telescope that one looks through. Eyepieces have three characteristics, which are: focal length, field-of-view (FOV) and eye-relief. These characteristics are determined by the shape and number of lenses in the eyepiece.

The focal length is written on the side or top of almost all modern eyepieces and is expressed in millimeters. The eyepiece focal length determines the telescope magnification and the telescope FOV. The ability to have different magnifications and different fields of view are two reasons why people often have several different eyepieces. Eyepieces should be stored in the kind of case like the one shown in Figure 4.12. Never place unprotected eyepieces in a pocket or purse.

Figure 4.12 An eyepiece storage box. Note that eyepieces are protected from being scratched. (Credit: Richard W. Schmude, Jr.)

A second important eyepiece characteristic is its apparent field-of-view (AFOV), which describes the amount of the eye's FOV that is covered by the eyepiece. As an example, if an eyepiece has a 55° AFOV, it covers about one-third of the eye's 170° FOV. One must remember that the eyepiece's AFOV is different from the telescope's AFOV. The telescope AFOV will depend on the magnification, eyepiece AFOV and characteristics of the telescope optics, and it will always be much smaller than the eyepiece AFOV.

The third eyepiece characteristic is eye-relief, which is the maximum distance between the eye and the eyepiece where the entire AFOV is visible. The smaller the eye-relief, the closer one has to hold his/her eyes to the eyepiece to see the entire field. If an eyepiece has an eye-relief of at least 15 mm, people who wear eyeglasses should be able to see the entire AFOV with their glasses on.

High quality eyepieces should yield excellent contrast and give images of stars as nearly pinpoints of light. Contrast is determined by the type of glass used in the eyepiece lenses along with the optical coatings, the quality of the lenses and the magnification. Old eyepieces may not have coatings on the lens-to-air surfaces and, as a result, the lenses will reflect more light resulting in a dimmer view. The quality and alignment of the lenses in the eyepieces determine how well one can focus. Many eyepieces will yield a sharp image of the target when it is centered in the FOV but a poor image when it is near the edge of the FOV. If one plans to study Uranus and Neptune, this would not be a serious problem; but if they wish to observe the moons of these planets, they should obtain an eyepiece that gives sharp images out to the edge.

Finding Uranus and Neptune

Finder charts for Uranus and Neptune are published each year in: *Sky and Telescope* magazine, *Astronomy* magazine, *The Observer's Handbook* (published by the Royal Astronomical Society of Canada), and *The Handbook of the British Astronomical Association*. These finder charts are helpful. In addition, the first two sources often list larger sky charts that can be used in locating the general location

Table 4.3. Selected comparison stars for visual and photoelectric photometry of Uranus for the 2008–2020 apparitions

Apparition	Suggested Comparison Stars for Uranus
2008–09	96-Aqr, HD221356, 14-Psc, λ-Psc
2009–10	HD221356, 13-Psc, 14-Psc, λ-Psc
2010–11	λ-Psc, 21-Psc, HD6, 14-Cet
2011–12	λ-Psc, HD6, 14-Cet
2012–13	14-Cet, δ-Psc
2013–14	60-Psc, δ-Psc, HD4928, ε-Psc, 73-Psc
2014–15	60-Psc, δ-Psc, HD4928, ε-Psc, 73-Psc
2015–16	ε-Psc, 73-Psc, 88-Psc, μ-Psc
2016–17	ε-Psc, 88-Psc, μ-Psc, π-Psc
2017–18	88-Psc, μ-Psc, π-Psc
2018–19	π-Psc, ν-Psc, HD11592, ι-Ari, 64-Cet, θ-Ari
2019–20	64-Cet, θ-Ari, ξ-Ari, $ξ^2$-Cet, σ-Ari
2020–21	64-Cet, θ-Ari, ξ-Ari, $ξ^2$-Cet, σ-Ari

of Uranus and Neptune. Star magnitudes are not given usually in finder charts; however, I have prepared tables listing comparison stars for both Uranus and Neptune. Table 4.3 lists selected comparison stars that can be used for Uranus, while Table 4.4 lists the Johnson B, V, R and I magnitudes along with visual magnitudes for the comparison stars. The visual magnitudes were computed from equation 4.1. Tables 4.5 and 4.6 contain the same data for comparison stars near Neptune. None of the stars listed in Tables 4.4 and 4.6 are listed as variable stars in the Millennium Star Atlas. One should follow the steps in Table 4.7 when using binoculars to locate Uranus and Neptune. The important thing to remember is to look for star patterns near Uranus which are the same as those in the finder chart. These star patterns serve to confirm the planet's location. Keep in mind that Uranus will appear like a star in binoculars.

It is more difficult to use a telescope than binoculars to find Uranus because the telescope may show things upside down, reversed or both upside down and reversed; furthermore, a telescope almost always has a smaller FOV than a pair of binoculars. Most finder charts are oriented for binoculars, which show things in the same way that our eyes see them. Therefore, before beginning a search, one should find out how his or her telescope and finder orients star patterns near the target.

The first step that I take in searching for a faint target like Uranus or Neptune is to locate it with binoculars. The binoculars reveal the pattern of stars near the target, and this helps me because I have a better idea of where to point the finderscope. Once I have pointed the finder at the target, I look through the telescope using low magnification. If I suspect the target is there, I look for other stars near it and compare them to the view through the finder. Once I am convinced that the stars match up, I often increase the magnification to confirm the presence of the target. Table 4.8 lists the steps needed to find Uranus or Neptune with a telescope.

People using an automated telescope should test it out on a familiar object like our Moon or Jupiter to make sure that it is operating properly before trying it on Uranus or Neptune. One should confirm the presence of the target by increasing the magnification and then lower the magnification and make the brightness estimate.

Table 4.4. Positions and magnitudes for the comparison stars for visual and photoelectric photometry of Uranus

Star	R.A. (2000) [c]	Dec. (2000) [c]	m_v [d]	B, V, R and I magnitudes
96-Aqr [a]	23 h 19 m 24 s	−5° 07′ 28″	5.6	5.94, 5.55, no R or I values
HD221356 [b]	23 h 31 m 32 s	−4° 05′ 14″	6.6	7.04, 6.496, no R or I values
13-Psc [b]	23 h 31 m 58 s	−1° 05′ 10″	6.6	7.57, 6.391, no R or I values
14-Psc [b]	23 h 34 m 09 s	−1° 14′ 51″	5.9	6.19, 5.890, no R or I values
λ-Psc [a]	23 h 42 m 03 s	1° 46′ 48″	4.5	4.71, 4.49, 4.30 and 4.20
21-Psc [b]	23 h 49 m 28 s	1° 04′ 34″	5.7	5.99, 5.763, no R or I values
HD6 [b]	0 h 05 m 04 s	−0° 30′ 09″	6.5	7.40, 6.296, no R or I values
14-Cet [b]	0 h 35 m 33 s	−0° 30′ 20″	6.0	6.37, 5.931, no R or I values
60-Psc [b]	0 h 47 m 24 s	6° 44′ 27″	6.2	6.92, 5.981, no R or I values
δ-Psc [b]	0 h 48 m 41 s	7° 35′ 06″	4.8	5.94, 4.426, 3.26 and 2.39
HD4928 [b]	0 h 51 m 18 s	3° 23′ 06″	6.6	7.43, 6.364, no R or I values
ε-Psc [b]	1 h 02 m 57 s	7° 53′ 24″	4.5	5.24, 4.273, 3.50 and 2.97
73-Psc [b]	1 h 04 m 53 s	5° 39′ 23″	6.3	7.52, 6.007, no R or I values
88-Psc [b]	1 h 14 m 42 s	6° 59′ 44″	6.3	7.11, 6.026, no R or I values
μ-Psc [b]	1 h 30 m 11 s	6° 08′ 38″	5.1	6.22, 4.842, 3.77, 3.03
π-Psc [b]	1 h 37 m 06 s	12° 08′ 30″	5.7	5.88, 5.56, no R or I values
ν-Psc [a]	1 h 41 m 26 s	5° 29′ 15″	–	5.80, 4.44, 3.38, 2.67
HD11592 [a]	1 h 53 m 58 s	10° 37′ 05″	6.9	7.24, 6.78, no R or I values
ι-Ari [a]	1 h 57 m 21 s	17° 49′ 03″	5.3	6.02, 5.10, no R or I values
64-Cet [b]	2 h 11 m 21 s	8° 34′ 12″	5.7	6.21, 5.623, no R or I values
θ-Ari [b]	2 h 18 m 08 s	19° 54′ 04″	5.6	5.63, 5.620, no R or I values
ξ-Ari [b]	2 h 24 m 49 s	10° 36′ 38″	5.4	5.37, 5.460, no R or I values
ξ²-Cet [b]	2 h 28 m 10 s	8° 27′ 36″	–	4.22, 4.282, 4.24, 4.28
σ-Ari [b]	2 h 51 m 30 s	15° 04′ 55″	5.5	5.40, 5.480, no R or I values

[a] The B, V, R and I magnitudes are from Iriarte et al 1965 and the magnitudes for stars with just B and V magnitudes are from Hirshfeld et al 1991.

[b] V filter magnitudes are from the General Catalogue of Photometric Data (at http://Vizier.cfa.-harvard.edu/vis-bin/vizier-3). I computed the B, R and I magnitudes from the V magnitudes listed along with the color indexes in Iriarte et al (1965). For stars with just B and V values, I computed the B magnitudes by using the V magnitudes along with the B–V value listed in Hirshfeld and Sinnott, 1991.

[c] All position values are from Hirshfeld et al (1991).

[d] I computed all visual magnitudes from the B and V magnitudes along with equation 4.1. with b = 0.21

The three ways of measuring brightness are: (1) visual, (2) photoelectric photometry and (3) CCD photometry. The second and third methods are discussed in the next chapter. We will discuss visual brightness measurements in the following section.

Visual Brightness Estimates

The brightness of stars and planets are arranged on the magnitude scale. The brighter an object, the smaller (or more negative) is its magnitude. As an example, the Full Moon is between magnitude −12.5 and −13.1, Polaris (the North Star) is

Table 4.5. Selected comparison stars for visual and photoelectric photometry for Neptune for the 2008–2020 apparitions

Apparition	Suggested Comparison Stars for Neptune
2008–09	HD205130, HD207439, HD208704, ι-Aqr
2009–10	HD205130, HD207439, HD208704, ι-Aqr
2010–11	HD205130, HD208704, ι-Aqr, 36-Aqr, HD210752
2011–12	HD208704, ι-Aqr, 36-Aqr, HD210752, HD211234, σ-Aqr
2012–13	HD208704, ι-Aqr, 36-Aqr, HD210752, HD211234, σ-Aqr
2013–14	ι-Aqr, 36-Aqr, 38-Aqr, HD210752, HD211234, σ-Aqr
2014–15	HD211380, σ-Aqr, HD214183, 64-Aqr
2015–16	σ-Aqr, HD214183, 64-Aqr, HD214722, 78-Aqr
2016–17	σ-Aqr, HD214183, 64-Aqr, HD214722, 78-Aqr
2017–18	HD214183, HD217580, HD217877
2018–19	HD217580, HD217877, HD220172, HD220339
2019–20	HD217580, HD217877, HD220172, HD220339
2020–21	HD217877, HD220339, HD221777

Table 4.6. Positions and magnitudes for the comparison stars for visual and photoelectric photometry of Neptune

Star	RA (2000) [c]	Dec (2000) [c]	m_v [d]	B, V, R and I magnitudes
HD205130 [a]	21 h 33 m 35 s	−9° 39′ 38″	7.9	7.88, 7.88, no R or I values
HD207439 [a]	21 h 49 m 20 s	−18° 23′ 11″	7.6	7.9, 7.56, no R or I values
HD208704 [a]	21 h 58 m 24 s	−12° 39′ 53″	7.3	7.83, 7.21, no R or I values
ι-Aqr [b]	22 h 06 m 26 s	−13° 52′ 11″	–	4.20, 4.266, 4.31, 4.40
36-Aqr [a]	22 h 09 m 27 s	−8° 11′ 09″	7.2	7.97, 7.00, no R or I values
38-Aqr [b]	22 h 10 m 38 s	−11° 33′ 54″	–	5.31, 5.431, no R or I values
HD210752 [a]	22 h 12 m 44 s	−6° 28′ 08″	7.5	7.92, 7.40, no R or I values
HD211234 [a]	22 h 16 m 01 s	−14° 26′ 09″	8.1	9.08, 7.84, no R or I values
HD211380 [a]	22 h 16 m 57 s	−14° 39′ 24″	7.3	7.61, 7.17, no R or I values
σ-Aqr [a]	22 h 30 m 39 s	−10° 40′ 41″	–	4.76, 4.82, 4.83, 4.87
HD214183 [a]	22 h 36 m 40 s	0° 21′ 51″	7.8	7.94, 7.81, no R or I values
64-Aqr [a]	22 h 39 m 16 s	−10° 01′ 40″	7.3	7.60, 7.16, no R or I values
HD214722 [a]	22 h 40 m 12 s	−6° 32′ 05″	7.3	7.7, 7.13, no R or I values
78-Aqr [b]	22 h 54 m 34 s	−7° 12′ 17″	–	7.46, 6.183, no R or I values
HD217580 [a]	23 h 01 m 52 s	−3° 50′ 56″	7.7	8.41, 7.46, no R or I values
HD217877 [a]	23 h 03 m 57 s	−4° 47′ 43″	6.8	7.26, 6.68, no R or I values
HD220172 [a]	23 h 21 m 51 s	−9° 45′ 41″	7.6	7.49, 7.67, no R or I values
HD220339 [a]	23 h 23 m 05 s	−10° 45′ 52″	8.0	8.69, 7.80, no R or I values
HD221777 [a]	23 h 35 m 04 s	−7° 40′ 37″	7.6	8.62, 7.32, no R or I values

[a] B and V magnitude values are from Hirshfeld et al (1991).
[b] The V filter magnitude values are from the General Catalog of Photometry Data. The B, R, and I magnitudes were computed from the color indexes reported by Iriarte et al (1965). I computed the B filter magnitudes for stars with just B and V magnitudes from the V filter magnitudes along with the B–V values reported in Hirshfeld et al (1991).
[c] All position measurements are from Hirshfeld et al (1991).
[d] I computed the visual magnitudes from the B and V magnitudes along with equation 4.1. with b = 0.21.

Table 4.7. Steps to follow when locating Uranus or Neptune with binoculars

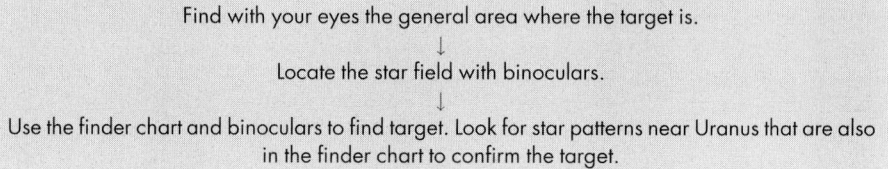

Find with your eyes the general area where the target is.
↓
Locate the star field with binoculars.
↓
Use the finder chart and binoculars to find target. Look for star patterns near Uranus that are also in the finder chart to confirm the target.

Table 4.8. Steps to follow when locating Uranus or Neptune with a telescope

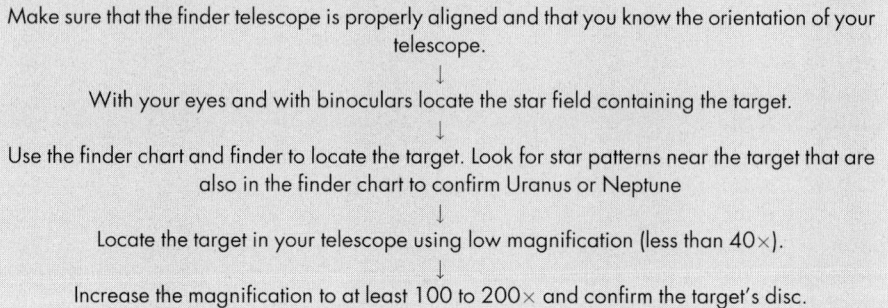

Make sure that the finder telescope is properly aligned and that you know the orientation of your telescope.
↓
With your eyes and with binoculars locate the star field containing the target.
↓
Use the finder chart and finder to locate the target. Look for star patterns near the target that are also in the finder chart to confirm Uranus or Neptune
↓
Locate the target in your telescope using low magnification (less than 40×).
↓
Increase the magnification to at least 100 to 200× and confirm the target's disc.

magnitude 2.0 and Uranus is magnitude ∼5.8. Compared to Uranus, Polaris is much brighter and its magnitude value is smaller. Compared to the North Star, the Full Moon is much brighter and its magnitude value is smaller (negative).

There are several limitations to the visual method of estimating magnitudes, which are:

(1) Accuracy is limited to ∼0.2 magnitudes,
(2) Different sensitivity to different color,
(3) Atmospheric extinction and
(4) Different background light.

These limitations are discussed briefly below.

Our eyes do not estimate quantitative brightness values very well; hence, visual magnitudes are only accurate to ∼0.2 magnitudes. If the comparison star is red and the target is green, accuracy will be lost because the eye is more sensitive to green than to red light. This error can be reduced by converting V filter magnitudes to visual magnitudes. If the comparison star is at a much lower altitude than the target, the brightness of the target will be incorrect. This is because the star light will be dimmed more by the atmosphere than the light coming from the target (which is higher in the sky). The best way of avoiding extinction errors is to use comparison stars with nearly the same altitude as the target. The background can affect the brightness of either the target or the comparison star. If possible, visual magnitude estimates should not be made when a bright Moon is present and should be made in a location where there is little scattered light. Light pollution, airglow, the zodiacal light, the Gegenschein, moonlight or the diffuse light from the galaxy can all affect the background light. In the next three paragraphs, I will describe airglow, the zodiacal light and the Gegenschein.

Airglow is a low-level brightening in the night sky that is caused by solar activity. Airglow gets brighter near solar maximum. Alan MacRobert of *Sky & Telescope* magazine reported that airglow was brighter at a dark site in Maine than at a similar site in Texas. This is probably due to Maine being closer to the magnetic pole than Texas. Photometric measurements made by astronomer Kevin Krisciunas show that the sky brightened a little in 1989 to 1992 due to solar maximum and from volcanic aerosols. Volcanic aerosols should not be a big factor for observing the remote planets since they are distributed uniformly across the sky

The zodiacal light is caused by the forward scattering of sunlight by interplanetary dust. It appears as a faint glowing oval centered on the Sun; half of the oval is visible before sunrise and the other half is visible after sunset. This dim light is about as bright as the Milky Way Galaxy. When Uranus or Neptune is within two months of conjunction, they will lie within a brighter portion of the zodiacal light and thus the sky background will be brighter. This may be a problem if either the comparison star or the target is in a deeper layer of the zodiacal light. If one plans to study these planets within two months of conjunction then he/she should choose comparison stars that are as close to the target as possible.

The Gegenschein is a faint, dimly lit oval that is in the opposite direction of the Sun. When Uranus or Neptune is within ten days of opposition, it will lie within the Gegenschein and the sky background will be higher than at other times. Since there is a brightness gradient near the edge of this feature, it can be a problem for photoelectric photometry but it should not be one for visual magnitude studies.

Before I make a magnitude estimate, I allow my eyes to become at least somewhat dark-adapted. On many occasions, I was unable to see Neptune through my 15×70 binoculars when I first stepped into the dark because my eyes were not fully dark-adapted. Rather than give up, I remained outside and took a few minutes to admire the dark sky while my eyes became dark-adapted. After four to five minutes, I was able to see Neptune and a fainter comparison star through my binoculars and was thus able to make a magnitude estimate of that planet.

In the visual magnitude method, one first finds the target and at least two nearby comparison stars of known magnitude. One star must be a little brighter and the other one must be a little dimmer than the target. The range between the comparison stars should be 1.0 magnitude or less. If the magnitude range exceeds this amount, then systematic errors in excess of 0.1 magnitudes may creep into the results. This is discussed more in a later section. The comparison stars must be close to the same altitude as the target so that extinction errors are kept to a minimum. When making the magnitude estimate, one should look quickly at the target and then at each of the comparison stars. I prefer to bounce back several times between the target and each comparison star. One should not stare at any of the objects for more than a second but instead glance at each one and rapidly move to the next one. After four or five cycles, one can estimate the brightness of the target by deciding its brightness in terms of the two stars. A mathematical ratio works best for me. Essentially, I estimate the brightness difference between the bright star and the target in terms of the difference between the target and the dim star. With this ratio, I compute the magnitude.

For example, let's say that Uranus is dimmer than a 5.6 magnitude star but is brighter than a 6.3 magnitude star; furthermore, the difference between Uranus and the dim star is twice the difference between the bright star and Uranus. By making this estimate, one breaks down the difference between the two comparison stars into three steps as illustrated in the sequence:

5.6 magnitude star

↓ (one step)

Uranus is one step fainter than the 5.6 magnitude star

↓ (one step)

↓ (one step)

The 6.3 magnitude star is two steps fainter than Uranus

The magnitude of Uranus is computed as:

$$(6.3 - 5.6)/3 \text{steps} = 0.23 \text{ magnitude/step}$$

$$5.6 + (0.23 \text{ magnitude/step}) \times 1 \text{ step} = 5.6 + 0.23 = 5.83 \sim 5.8.$$

The magnitude is 5.8 because the typical uncertainty for this kind of estimate is ~0.2 magnitudes.

Here is an example: Joe observes comparison stars of magnitude 7.4 and 8.2 and notes that Neptune is a little fainter than the 7.4 magnitude star but is a lot brighter than the 8.2 magnitude star. He feels that the difference between Neptune and the 8.2 magnitude star is three times the difference between that planet and the 7.4 magnitude star. What is Neptune's brightness?

For the solution, follow the chart below:

7.4 magnitude star

↓ (one step)

Neptune is one step fainter than the 7.4 magnitude star

↓ (one step)

↓ (one step)

↓ (one step)

The 8.2 magnitude star is three steps fainter than Neptune

The magnitude of Neptune is computed as:

$$(8.2 - 7.4)/4 \text{ steps} = 0.2 \text{ magnitudes/step}$$

$$\text{Neptune's brightness} = 7.4 + 0.2 \text{ magnitude/step} \times 1 \text{ step} = 7.4 + 0.2 = 7.6$$

Experiments

Here are two experiments to check the method just discussed. In the first experiment, we use two stars, Alpha-Lyrae ($m_v = 0.00$) and Theta-Herculis ($m_v = 4.14$), to estimate the magnitudes of ten stars of known brightness. The ten stars have brightness values between those of the comparison stars. We can make seven sets of visual magnitude estimates for each of these stars and compute average magnitudes values. In all cases, we can compute the literature values of m_v from Johnson V filter values using equation 4.1. and b = 0.21. All estimated magnitudes are fainter than the literature values. The results are shown in Figure 4.13. The discrepancies tend to be greater for stars with a magnitude of 2.0 to 2.4 compared to stars that are closer in brightness to one of the comparison stars. The average brightness discrepancy is 0.41 magnitudes. Much of this discrepancy is undoubtedly due to the wide range of comparison star magnitudes that prevented me from properly constraining the star magnitudes.

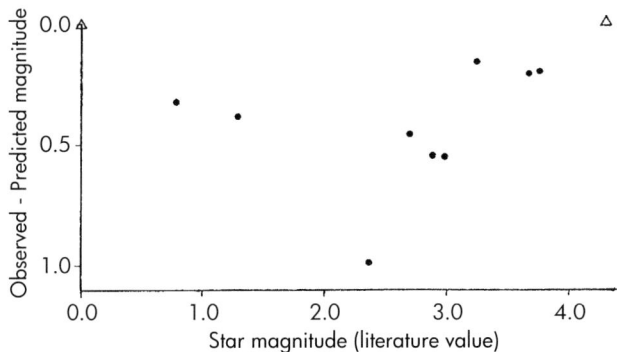

Figure 4.13 The writer estimated the brightness of several stars (of known brightness) in terms of Alpha-Lyrae and Theta-Herculis using the visual method. His goal was to see how accurate the visual estimates were to the literature values by using comparison stars with a brightness difference of over four magnitudes. The observed minus predicted (or literature) magnitude is plotted against the literature magnitude. In all cases, the observed magnitude was fainter than the literature brightness. The triangles are the two comparison star magnitude differences (0.0) and the filled circles are the observed minus predicted magnitudes of the stars studied. (Credit: Richard W. Schmude, Jr.)

Figure 4.14 The writer estimated the brightness of Mu-Pegasi and Iota-Pegasi in terms of Eta-Pegasi and Lambda-Pegasi using the visual method. His goal was to see how accurate the visual estimates were to the literature values by using comparison stars with a brightness difference of about one magnitude. The observed minus predicted (or literature) magnitude is plotted against the literature magnitude. In both cases, the observed magnitude was fainter than the literature brightness; however, the differences are much less than in the previous figure because the two comparison stars were just over one magnitude apart instead of being over four magnitudes apart. The triangles are the two comparison star magnitude differences (0.0) and the filled circles are the observed minus predicted magnitudes of the stars studied. (Credit: Richard W. Schmude, Jr.)

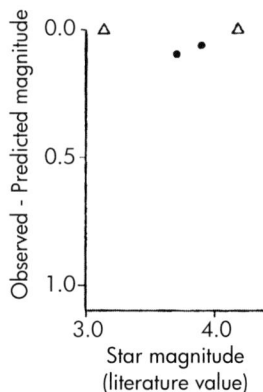

In a second experiment, I repeated the procedure in the first experiment except that I changed comparison stars. See Figure 4.14. I chose comparison stars that were much closer in brightness (Eta-Pegasi, $m_v = 3.13$ and Lambda-Pegasi, $m_v = 4.17$). I estimated the brightness of two stars with brightness values between the comparison stars four or five times each and then computed average magnitudes. The estimated magnitudes in this experiment were much closer to the literature values. The average error was a more acceptable 0.08 magnitudes. In conclusion, one should use comparison stars that are as close in brightness as possible to the target when employing the visual brightness method.

Color

Uranus, Neptune, and Pluto all have color; however, before their colors can be understood, the reader must appreciate what color is and the terms related to it. In this paragraph, we will describe the important characteristics of color. There are three

Table 4.9. Light wavelengths for different colors based on estimates made by the author with a calibrated spectroscope

Color	Wavelength Range (nanometers)
Red	640 to 710
Orange	595 to 640
Yellow	580 to 595
Green	505 to 580
Blue	450 to 505
Violet	410 to 450

dimensions of color; these are hue, saturation, and intensity. Hue is the dominant wavelength of the color. The wavelengths for different colors according to the writer's eyes are listed in Table 4.9. White, gray, and black do not have any hue. The saturation refers to how much white is mixed in with the dominant wavelength; in other words, saturation expresses the degree to which color departs from white and approaches a pure color. Red laser light is an example of light with a pure color and high saturation. Beams of equally intense red, blue and green light all pointed on the same area will produce a nearly white color with a lower saturation than the original three colors. The third dimension of color – the intensity – describes the amount of light entering the eye. The light intensity can affect the perceived color of an object.

The amount of each wavelength of light reflected (or emitted) by an object will determine the three dimensions of its color. A common fluorescent light appears white, but it gives off intense red, green, and indigo light along with small amounts of yellow, orange and blue-green light. Uranus on the other hand, reflects lots of green light along with smaller amounts of blue, yellow, orange and red light. The net result of this mixture is a faint greenish color, which is fairly close to white, and is far from a spectrally pure green color. The reason for this is that much of the blue, green, and red light reflected by Uranus is mixed to produce white light, and this, combined with additional green light, produces a green-white color. This is also why most red stars have a fainter red color than that of red traffic signals.

In order for the eye to detect color, a certain amount of light must enter it. For example, I am able to see a blue-white color for the star Vega ($m_v = 0.0$) and an orange-white color for the star Betelgeuse ($m_v = 0.3$) with the unaided eye. The situation is different for the star Polaris. At magnitude $m_2 = 2$, that star does not give off enough light for me to see any color. The only way that I can see Polaris's color is to increase the amount of light reaching my eyes. Through a pair of 70-mm binoculars, the yellow-white color becomes obvious to me.

If too much light reaches the eyes, the cones approach maximum stimulation and the color purity of the object approaches white. A telescope gathers light and, as a result, it causes more light to reach the cones in a person's eye. Therefore, a magnitude 0 star may appear to have a reddish color with the unaided-eye, but through a 0.3 m (12 inch) telescope, this star will appear white with a slight reddish hue.

At magnitude ~5.8, Uranus is too dim to show color. If one uses 12×70 binoculars to observe that planet, its brightness increases by a factor of ~100 and, as a result, there should be enough light for the eye to detect color. The light-gathering power of binoculars or a telescope affects how we perceive the color of Uranus and Neptune. I have observed Uranus with telescopes ranging in diameter from 4 to 30 inches (10 to 76 cm). The general trend that I have seen is that the

Table 4.10. Colors of Uranus observed by the writer for different telescope apertures

Aperture (type)	Magnification	Color
6 inch (Reflector)	30×, 190×	Bluish
10 inch (Reflector)	300×	Yellow-green center and Blue-green limb
14 inch (Schmidt-Cassegrain)	530×	Greenish hue
18 inch (Reflector)	460×	Pale sea-green
28 inch (Binocular reflector)	490×	Slight greenish
30 inch (Reflector)	230×, 370×	Washed-out blue-green

saturation of Uranus' color diminishes with increasing telescope aperture; that is, it appears whiter through large telescopes compared to smaller ones.

People generally report higher color saturation for Neptune than for Uranus with large telescopes. For example I noted a "washed-out blue-green" color for Uranus but a "bluish" color for Neptune with a 0.76 m (30 inch) Newtonian on July 19, 1992. As a second example, Norman Boisclair reported a "pale green" color for Uranus and a "very deep blue-gray" color for Neptune with a 50.8 cm (20 inch) Newtonian under excellent viewing conditions on October 4, 2005. Color perception through a telescope is undoubtedly complex. Part of the reason for the difference may be due possibly to the fact that Uranus is brighter than Neptune and the observer's eye is closer to maximum saturation for Uranus compared to Neptune. Much of the reason for the color difference between Uranus and Neptune, however, is that these two planets have different colors.

My color estimates with various telescope diameters are summarized in Table 4.10. The trend in Table 4.10 is consistent with color estimates made by ALPO members dating back to the early 1990 s.

If one wants to carry out color studies, they should follow a few rules. One of these is that the observer must use a consistent telescope diameter and magnification for color studies. Uranus and Neptune require minimum apertures of three and four inches while Pluto requires a diameter of ~1.5 meters (~60 inches). One must also carry out color studies when these objects are at least 30° above the horizon. The sky must also be clear with little or no haze present. Since different people see color differently, the most valuable color data are collected by the same individual over a long period of time. The individual should concentrate on any change(s) in the hue or purity of color.

Observing with Medium-Sized Telescopes

People have successfully measured the brightness of Uranus, Neptune, and Pluto with medium-sized telescopes (telescopes with diameters between 0.1 and 0.25 m or 4–10 inches). Doug West and Frank Melillo have also measured the light spectrum of Uranus with a medium-sized telescope. In this chapter, we will describe the different kinds of observations that one can make of the outer planets with a medium-sized telescope. This chapter starts out with introductory information, which is followed by discussions on filters, photometers, and cameras. A detailed example of a photometric measurement is presented. The chapter ends with discussions on imaging Uranus's light spectrum and making visual magnitude measurements of Pluto.

In most of this chapter, we will deal with making observations with equipment attached to the telescope. In many cases, this equipment may weigh several pounds. As a result, one should have a mount which will be able to hold the weight of the equipment and of any necessary counterweights along with the telescope.

One should choose a convenient site to build his or her observatory. This is more important than choosing a dark site. Uranus and Neptune are relatively bright objects, and moderate light pollution should not be a problem. Try to avoid carrying out measurements when there is a lot of scattered light, such as on full-moon or gibbous-moon nights. One should review Table 5.1 before building an observatory.

One can do excellent remote planets' work with a portable telescope. The telescope should not be too big or too heavy. One should also be able to fit the telescope in his or her vehicle. The Schmidt-Cassegrain telescope is especially easy to transport.

Two important accessories that one should take are a dew-shield and a tarp or blanket. A Schmidt-Cassegrain must have a dew-shield because of the location of the corrector plate. Dew can condense on it and ruin the view. A dew-shield for the finder may also be a good thing to bring along. If there are lots of knobs and bolts to be screwed into the telescope or mount, it would be a good idea to lay a blanket or tarp down before setting up. In this way, if a small part is dropped, then it will not get lost in any grass or vegetation.

Photoelectric Equipment

A filter is a device that transmits certain wavelengths of light and blocks out other wavelengths of light. Different filters can block out different types of light. One characteristic of a filter is transmission, which defines the amount of light that it

R.W. Schmude, Jr., *Uranus, Neptune, and Pluto and How to Observe Them*, DOI: 10.1007/978-0-387-76602-7_5, © Springer Science+Business Media, LLC 2008

Table 5.1. Observatory checklist

Stage	Comments
Planning	Check local zoning regulations before building an observatory.
Planning	Obtain a building permit if necessary.
Planning	Avoid using concrete blocks or bricks. They absorb heat and can impair the seeing.
Planning	Allow for a larger telescope or for guests. Insure that the observatory is of sufficient size.
Building	Make sure that the telescope pier is not touching the observatory floor; otherwise the telescope will shake every time that somebody is moving inside of the observatory.
Building	Install a security alarm; a dog may also be helpful.
Building	Design the roof so that snow does not accumulate.
After completion	Check periodically for unwanted pests such as wasps, bees, and snakes.

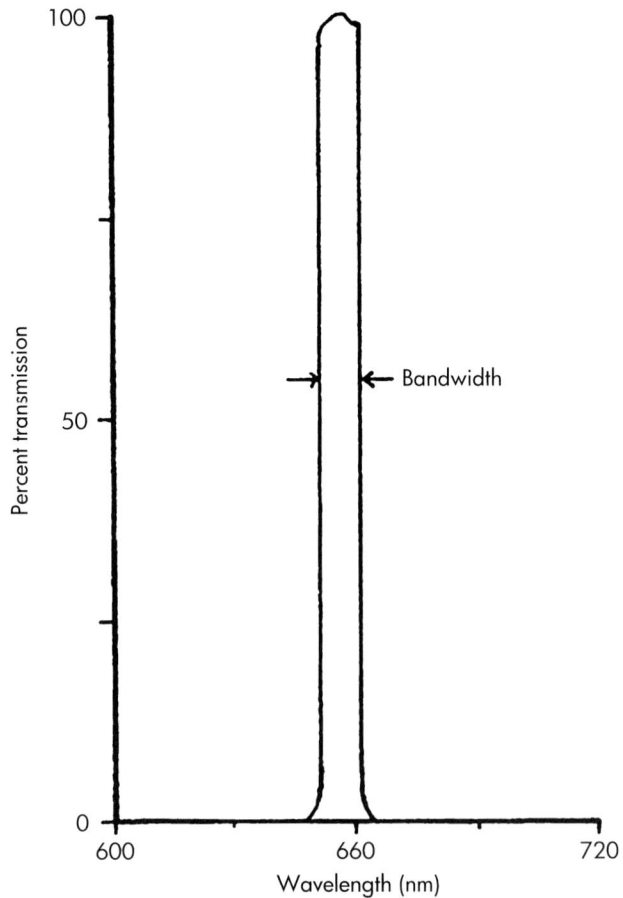

Figure 5.1. A graph of percent transmission versus wavelength for a filter having a peak transmission at a wavelength of 656 nm and a bandpass or bandwidth of 10 nm. (Credit: Richard W. Schmude, Jr.)

allows through. The percent transmission is a measure of the percentage of light that passes through. As an example, clear glass has a high percentage transmission of visible light passing through whereas tinted glass has a lower percentage transmission. The bandpass or bandwidth defines the wavelength range that a filter transmits.

A hydrogen-alpha filter, with a peak transmission for light having a wavelength of 656 nm with a bandpass of 10 nm allows in light with a wavelength range of 651–661 nm. See Figure 5.1. Most of the light with a wavelength of 655 nm will pass through this filter, but light with a wavelength of 630 nm will be blocked. A more precise way of describing the bandpass of a filter is the Full-Width-at-Half-Transmission (FWHT). This describes the wavelength range at which the transmission is at least half of the peak transmission. Figure 5.2 illustrates a FWHT of 497–587 nm. One must realize that some light outside of this range will still pass through; however, the transmission will be less than 50% of the peak value.

There are three categories of filters – broad-band, intermediate-band and narrow-band. Broad-band filters include those transformed to the Johnson B, V, R and I System; their bandpasses are 90–220 nm. The most popular intermediate-band filters are the Stromgren u, v, b, and y filters, which correspond to peak wavelengths of: 342, 410, 470, and 550 nm, respectively. These filters have bandpasses of 16–25 nm. Since these have smaller bandpasses, they let in less light than the broad-band filters, and may require more sensitive equipment. Narrow-band filters have bandpasses of less than 10 nm. The methane band filter that R. B. Minton used in studying Jupiter had a peak transmission value of 884 nm and

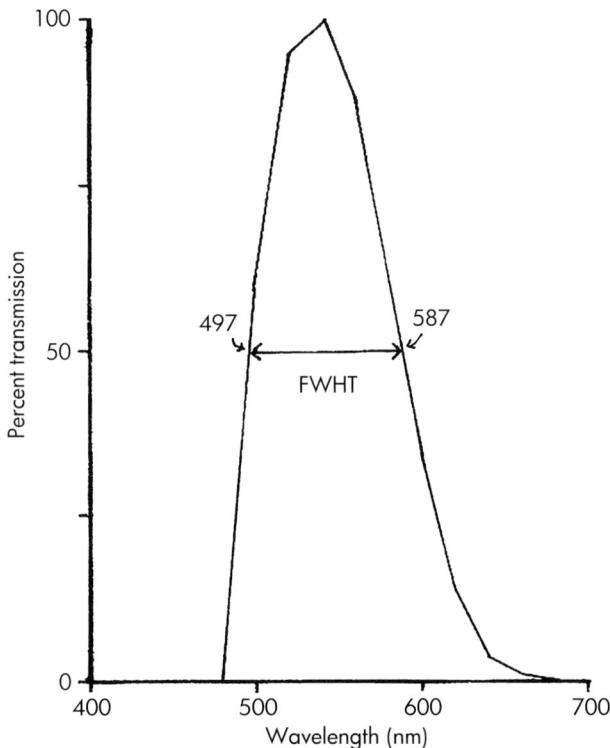

Figure 5.2. A graph of percent transmission versus wavelength for a filter. This graph illustrates the Full-Width at Half Transmission (FWHT) for a filter. (Credit: Richard W. Schmude, Jr.)

Table 5.2. Characteristics of a few commercially available filters that can be used for photometry

Filter	Type	Approximate Bandpass	Approximate Peak Transmission
B	Broad-band	90 nm	420 nm
V	Broad-band	90 nm	540 nm
R	Broad-band	200 nm	700 nm
I	Broad-band	220 nm	860 nm
J	Broad-band	130 nm	1270 nm
H	Broad-band	320 nm	1665 nm
u	Intermediate-band	25 nm	342 nm
v	Intermediate-band	16 nm	410 nm
b	Intermediate-band	19 nm	470 nm
y	Intermediate band	24 nm	550 nm

a FWHT value of 7 nm. This is an example of a narrow-band filter. Table 5.2 summarizes a few commercially available filters.

The temperature can affect the transmission characteristics of filters. For this reason, one should measure transformation coefficients near 10°C (50°F) if they plan to make measurements in the range of –5° to 25°C.

Figure 5.3 shows the transmission characteristics for one batch of filters close to the Johnson B, V, R and I system; the response function is proportional to the percent transmission. It is important to realize that manufacturers can change how they make filters, and this could cause transmission characteristics to change.

Filters should be protected from dust, loss, scratches and humidity. I store my glass filters inside of a photometer storage box. See Figure 5.4. The Stromgren filters require protection from high levels of humidity. When these filters are not being used, they should be stored in a container with desiccant. This keeps the air inside dry.

One can measure the brightness of a target with different color filters and thus obtain quantitative color information. In order to do this, one must calibrate their measurements to a standard system. Let me explain. Essentially, light interacts with at least four surfaces in a photometric system. These surfaces are the filter, the telescope optics, the photometer optics and the light detector. These surfaces together define the system response. Each system has its own sensitivity to each wavelength of light along with its own FWHT. To make matters more complicated, each star and planet gives off it own pattern of light called a spectrum.

The differences in system response and star/planet spectrum can be a source of systematic error, especially if a broad-band filter is used. Simply put, two observers with different systems may come up with two different brightness measurements of the same object. Broad-band filters are particularly prone to this error; hence, transformation corrections must be made. These corrections are described later. Intermediate-and narrow-band filters are less prone to this problem and, in many cases, transformation corrections are negligible.

The reader must remember that there is no such thing as a standard Johnson V filter. One must calibrate his or her equipment before reporting magnitudes in the Johnson V system. This calibration is the measurement of transformation coefficients, which is described in the Appendix. (Throughout this Book, I have avoided

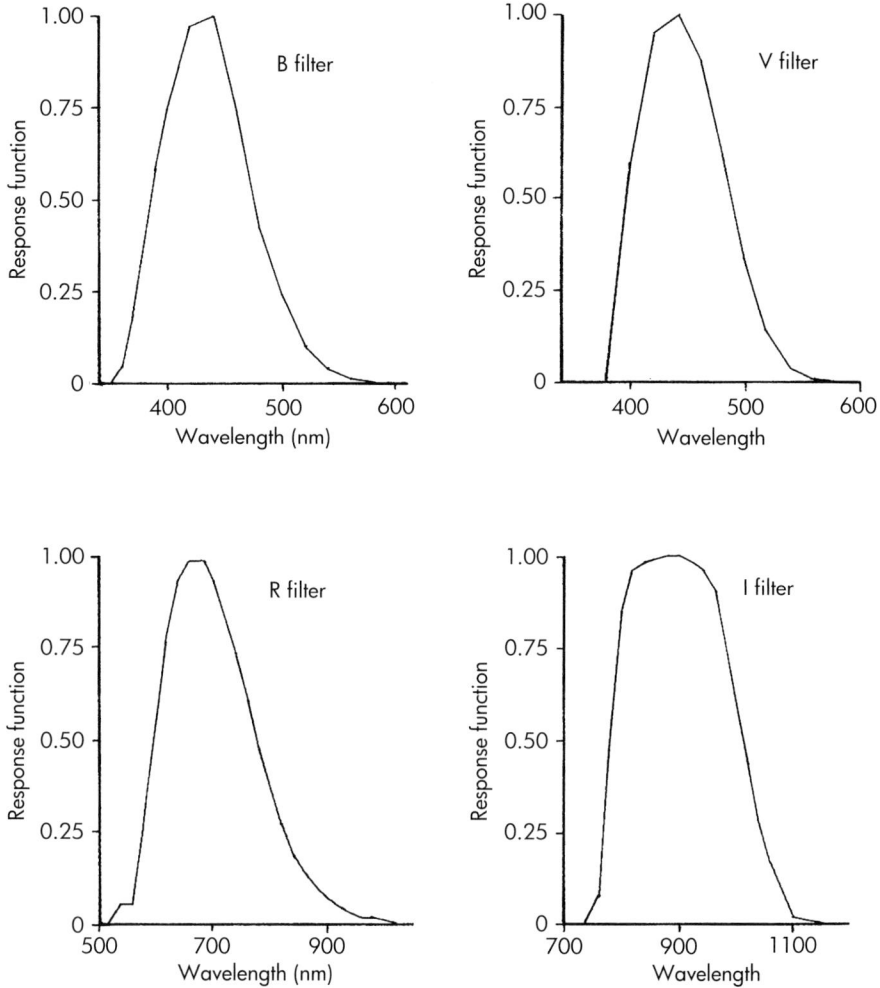

Figure 5.3. Transmission characteristics for a batch of filters close to the Johnson B, V, R and I band passes. The response function is proportional to the percent transmission. (Credit: Optec Inc. values are from the technical Manual for the SSP-3 solid state photometer. Drawings were made by Richard W. Schmude, Jr.)

the phrase "Johnson V filter" and have instead used the phrase "a filter transformed to the Johnson V system".)

The photoelectric photometer collects light and converts it into an electrical signal, which goes to either a digital readout or to the computer. One commercially available photometer is shown in Figure 5.5. This instrument has a flip mirror, which allows the observer to both focus his/her telescope and to center the object on the detector. The basic steps in the production of a signal in the SSP-3 and SSP-5 photometers are shown in Figure 5.6.

A good photometer should have four characteristics, namely, (1) linearity, (2) stability, (3) sensitivity and (4) durability, and it should be easy to use. Linearity means that the amount of light coming in should be directly proportional

Figure 5.4. Storage box for a photometer and filters. The box protects the filters from scratches and damage. (Credit: Richard W. Schmude, Jr.)

Figure 5.5. An SSP-3 photometer. Note that it fits into the 1.25 inch eyepiece hole of a telescope. (Credit: Richard W. Schmude, Jr.)

to the readout number. Therefore, if 800 photons strike the detector, the reading should be 20 times greater than if just 40 photons struck it (after the dark current is subtracted).

The second characteristic is stability, which means that the sensitivity of the photometer remains constant over time. If the sensitivity changes during an observing run, serious systematic error will be introduced into the measurements.

Sensitivity deals with how well the photometer measures faint sources of light. A reliable measurement will have a signal-to-noise ratio of at least 100. The signal is what the photometer measures and noise is random fluctuations. Some sources of noise are: dark current fluctuations, changes in the electronics and slight changes in atmospheric transparency and refraction. The telescope diameter, detector characteristics and the wavelength of light will all affect the sensitivity.

SSP-3 Photometer

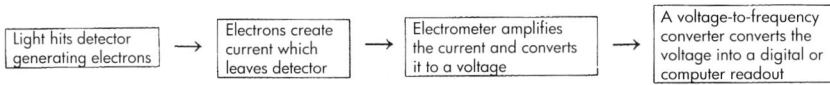

| Light hits detector generating electrons | → | Electrons create current which leaves detector | → | Electrometer amplifies the current and converts it to a voltage | → | A voltage-to-frequency converter converts the voltage into a digital or computer readout |

SSP-5 Photometer

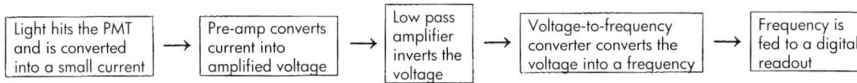

| Light hits the PMT and is converted into a small current | → | Pre-amp converts current into amplified voltage | → | Low pass amplifier inverts the voltage | → | Voltage-to-frequency converter converts the voltage into a frequency | → | Frequency is fed to a digital readout |

Figure 5.6. Steps in the generation of a signal in the SSP-3 (top) and SSP-5 (bottom) photometers. A PMT is a photomultiplier tube. (Credit: Richard W. Schmude, Jr.)

The fourth characteristic is durability. The photometer should last several years and be able to recover from minor observer mistakes or normal wear and tear. The photometer should be stored in a strong and durable case like the one in Figure 5.4.

A CCD camera contains an array of microscopic detectors (pixels) called a chip. Each pixel possesses linearity, stability, sensitivity and durability. Unlike the photometer, one can record an image and do photometry with it. One problem with CCD cameras, however, is that not all pixels in the chip have the same sensitivity. Therefore, one should take a flat-field image. This is an image taken of an area of equal brightness. Since not all pixels have an equal sensitivity, the flat-field image will not have a uniform intensity. One often uses an illuminated screen or the twilight sky to record it. A flat-field image should be made each time the camera is readjusted or when a different filter is used. In addition to a flat-field image, one should also make a dark-current image (often called a dark-frame). This image corrects for the fact that some pixels generate current from other sources besides the light falling on them. One takes a dark-current image by closing the shutter. After this, one subtracts it from the image. One must take the dark-current image at the same temperature as the images. If the camera temperature changes during an imaging sequence, the observer may have to take at least one additional dark-current image and use it in their processing routine.

The webcam is similar to a CCD camera except that it is not possible to make flat field and dark current corrections; therefore, these cameras should not be used for photometry. A second limitation of webcams is that it is not possible to have exposure times longer than about two seconds. In spite of these limitations, one can make good pictures of Uranus and Neptune with these devices.

A few people have been able to image the brighter planets with electronic cameras. Unless one is able to make flat-field and dark-current corrections to electronic camera images, one should not use them to make brightness measurements. Most people use a Barlow lens to increase their image size. Most electronic cameras also allow one to enlarge their images.

Medium-Sized Telescopes

Table 5.3. Equipment and useful projects

Equipment	Photometry	Imaging	Astrometry	Occultation/Transit
Photoelectric Photometer	Yes	No	No	Yes
CCD Camera	Yes	Yes	Yes	Yes
Webcam	No	Yes	Yes	Yes
Electronic Camera	No	Yes	Yes	Yes

Table 5.3 lists the various projects that one can do with different pieces of equipment. Keep in mind that improvements are always being made to electronic cameras and webcams. In the future it may be possible to do more with these instruments than what is listed in Table 5.3.

Photoelectric Photometry

Differential photometry is a technique whereby one measures the brightness difference between one object (the target) and another object (the comparison star). One objective of differential photometry is to measure the brightness of a target. When people measure the brightness through different color filters they obtain quantitative color information.

The brightness of a planet depends on at least five factors, namely, (1) the planet-Sun distance, (2) the planet-Earth distance, (3) the planet's sub-Earth latitude, (4) the solar phase angle and (5) the planet's albedo.

The first and second factors affect the brightness of Uranus and Neptune more than the other three factors during a year. During 2007, Uranus underwent a 0.22 magnitude brightness change as a result of changing distances, while the corresponding value for Neptune was 0.15 magnitudes. Pluto has a very elliptical orbit and, as a result, the changing distance will have different effect on its brightness in different years. In 2007, as a result of changing distances, Pluto was 0.15 magnitudes brighter in June than in December.

The planet's sub-Earth latitude – the latitude that appears at the center of the planet's disc as seen from the Earth – can affect the brightness in two ways. The first way is that the polar and equatorial regions on all three planets reflect different amounts of light. In fact, this is one of the things that astronomers are interested in determining when carrying out brightness measurements.

The sub-Earth latitude can also affect the brightness of a planet with a large polar flattening. Since Uranus has an ellipticity of 0.0229 and its polar or equatorial regions can face Earth, its brightness can change by 0.025 magnitudes. Essentially, when the sub-Earth latitude is $0°$, we see the true north-south dimension like what is shown in Figure 5.7. When the sub-Earth latitude is at $50°S$ as in Figure 5.8, we see a north-south dimension that is larger than the true dimension. This means that the planet will reflect more light and be brighter. For Uranus and Neptune, the positions and orientations of the moons and rings have little impact for visible light studies. Charon can cause a 0.15–0.20 magnitude brightness increase if it is in the photometer field-of-view. In cases where one is unable to separate Charon from Pluto, the magnitude should be reported for "Pluto + Charon".

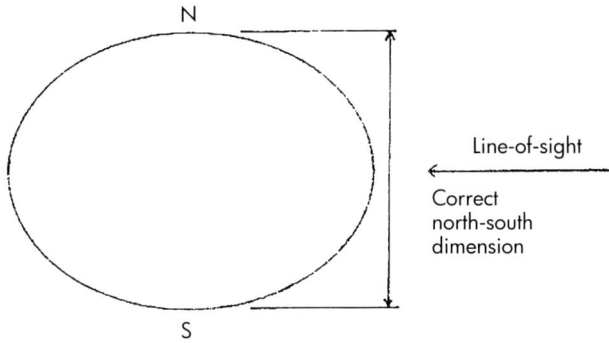

Figure 5.7. View of an oblate planet when the sub-Earth latitude is at the planet's equator. The sub-Earth latitude is the latitude at the center of the disc as seen from the Earth. Note that we see the true size of the planet's north-south dimension in this figure. (Credit: Richard W. Schmude, Jr.)

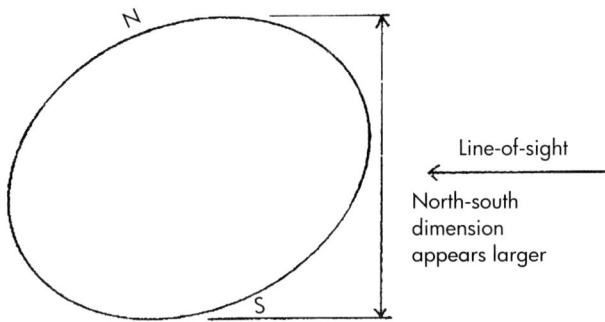

Figure 5.8. View of an oblate planet when the sub-Earth latitude is far from the planet's equator. In this case, an oblate planet's north-south dimension appears larger than what it is, causing the planet to reflect more light. (Credit: Richard W. Schmude, Jr.)

The change due to a planet's solar phase angle needs some explaining. When a planet is directly opposite from the sun, light is reflected straight back to the observer, but, as the planet moves away from this point, it becomes darker. This is due to the extra shading on the side not facing the Sun directly. This shading causes the planet to become dimmer. The change in brightness due to different solar phase angles (the fourth such factor) will be less than 0.01 magnitudes for Uranus and Neptune; hence, this can be neglected except in very precise work. During 2007, Pluto was 0.05 magnitudes brighter (through the V filter) in June at a low solar phase angle than in December when it was at a high solar phase angle.

The fifth factor which affects a planet's brightness is its albedo. The albedos of Uranus and Neptune can change by a few percent in visible light due to seasonal changes. Pluto, on the other hand, has bright and dark regions. As a result, when Pluto rotates, areas with different albedos come into view, causing the brightness to change.

The brightness of the moons of Uranus, Neptune and Pluto depend on the same five factors that determine a planet's brightness. The main difference for the moons, however, is the opposition surge. The large moons of Uranus and Pluto's largest moon, Charon, increase by a few tenths of a magnitude at opposition. The smaller moons probably have similar opposition surges. Triton, Neptune's largest moon, gets a few hundredths of a magnitude brighter at opposition.

Five changes that one can measure with photoelectric magnitude measurements of Uranus, Neptune and Pluto are: (1) color changes, (2) albedo changes caused by rotation, (3) albedo changes caused by cloud and haze development/dissipation, (4) albedo changes caused by changes in the sub-Earth latitude and (5) changes due to the changing solar phase angle.

The accuracy of a photometric measurement depends on the signal-to-noise ratio (SNR). Noise is random fluctuations in the data. The SNR is the signal divided by the noise. For example, an SNR = 100 means that the signal is 100 times greater than the noise. When taking measurements, the scientist strives to have as high of an SNR as possible. One can improve this quantity by taking additional measurements and taking an average. The SNR improves by a factor of $(N)^{1/2}$ where N is the number of measurements. For example, if the SNR of one measurement is 100, and if this measurement is taken four times the SNR of the average value is $100 \times (4)^{1/2} = 200$. The integration time is the length of time that the detector is exposed to the light source. The longer it is exposed, the greater will be the SNR. The SNR also increases with increasing telescope aperture. In Table 5.4, I have listed the faintest magnitude that one can measure, and obtain an SNR of 100 for various telescope apertures with a 90 second integration time for the SSP-3 photometer. Table 5.5 lists the same thing for the SSP-5 photometer. For example, the V filter magnitude of Neptune is around +7.7; hence, a 0.36 m telescope with an SSP-3 photometer is adequate to give an SNR close to 100 for a 90 second integration time. If one considers the B filter (Neptune B magnitude equals about 8.1) a 0.36 m telescope with the same photometer will not yield an SNR of 100 for a 90 second integration time. In this case, if one wanted an SNR of 100, he or she would have to use a longer integration time or take more measurements and compute an average.

In order to obtain an accuracy of 0.001 magnitudes, one must observe under excellent sky conditions (good seeing, transparency and probably be at an elevation of at least 1.0 km), have an SNR value of at least 1000 and know their comparison star magnitudes to an accuracy of 0.001 magnitudes. In addition, the comparison star should be constant to an accuracy of 0.001 magnitudes.

H. L. Johnson and co-workers compiled a list of stars with brightness values quoted to 0.001 magnitudes. Even our Sun undergoes brightness changes of up to 0.002 magnitudes. The problem of star variability is of special concern since magnitude changes of less than 0.01 magnitudes are often below the brightness

Table 5.4. Faintest magnitude that will yield an SNR = 100 for different telescope apertures with the SSP-3 photometer

Aperture Meters	(inches)	B, V and I filters	R filter	U filter
0.10	(4)	4.8	5.3	2.8
0.15	(6)	5.6	6.1	3.6
0.20	(8)	6.3	6.8	4.3
0.25	(10)	6.8	7.3	4.8
0.28	(11)	7.0	7.5	5.0
0.36	(14)	7.5	8.0	5.5
0.51	(20)	8.3	8.8	6.3
0.61	(24)	8.7	9.2	6.7
0.76	(30)	9.2	9.7	7.2
1.0	(40)	9.8	10.3	7.8

Table 5.5. Faintest magnitude that will yield an SNR = 100 for different telescope apertures with the SSP-5 photometer having the R6358 detector

Aperture Meters	(inches)	V filter	B filter
0.10	(4)	10.3	11.6
0.15	(6)	11.1	12.4
0.20	(8)	11.8	13.1
0.25	(10)	12.3	13.6
0.28	(11)	12.5	13.8
0.36	(14)	13.0	14.3
0.51	(20)	13.8	15.1
0.61	(24)	14.2	15.5
0.76	(30)	14.7	16.0
1.0	(40)	15.3	16.6

uncertainties in many star catalogs. Astronomers at the Lowell Observatory have attained an accuracy of 0.001 magnitudes. They use three comparison stars and also measure the brightness of comparison stars that they intend to use in future years. By compiling brightness data of future comparison stars, they can detect any variability before the star is used as a comparison object. Since three comparison stars are used, any variability can be detected, and furthermore, these astronomers are able to determine which of the comparison stars is variable. Potential comparison stars which are variable are not used as comparison stars in future studies.

I like to take at least three sets of measurements of the comparison star and target when making a brightness measurement. The measurements are taken in the sequence of CTCTCTC where C is a comparison star measurement and T is the target measurement. I will measure often a check star as well. Each C and T measurement consists of three 10-second sky brightness measurements and three 10-second measurements of the comparison star (or target). Three target magnitudes are computed from the sequence and an average brightness value is computed. Each value thus has 90 seconds of integration time for the target. The two advantages of taking three sets of T measurements are: (1) an average value of three sets of measurements is more reliable than a single set of measurements and (2) any changes in sky transparency can be detected. Obviously, data taken during highly variable sky transparency is not valid.

Great care should be taken when selecting a comparison star. (I prepared a short list of suitable comparison stars for visual and photoelectric photometry of Uranus and Neptune in the previous chapter.) One must make sure that the comparison star is not a variable. A variable star is one that changes in brightness. Two excellent Atlases that I use are *The Millennium Sky Atlas* edited by Sinnott and Perrymann and *Sky Catalog 2000.0 Volume 1 Stars to Magnitude 8.0* written by Hirshfeld, Sinnott and Ochsenbein. These Atlases will alert the reader if a star is variable. Secondly, the comparison star should be as close to the target as practicable. In this way, the extinction correction and any associated error will be small. The comparison star should have a B–V value of around 0.5 because this is close to the B–V value of Uranus and Neptune. This will insure that the color correction and any associated uncertainty will be small. Finally, the comparison star should be as close to the brightness of the target as possible. One problem with this last point, however, is that very few stars fainter than magnitude 6.0 have measured red and infrared magnitudes. In a perfect world, comparison stars would have the same color and brightness as the target and would also be very close to it. In the real world, compromises must be made. I feel that the most critical factor is distance – the comparison star should be close to the target. This will reduce errors due to atmospheric extinction.

The heart of the photoelectric photometer is its detector. Table 5.6 lists a few commercially available photoelectric photometers from Optec Inc. along with their detectors, wavelength ranges and magnitude limits. I have used the SSP-3 extensively for making magnitude measurements of Uranus and Neptune. Figure 5.5 shows the SSP-3 photometer attached to a telescope.

There are a few things that one should do before making magnitude measurements. The first thing is to check the skies; they must be transparent and nearly haze-free. Even very thin clouds can be a problem; furthermore, the northern lights can pose a problem. If the skies are clear, one places the photometer securely into the 1.25 inch eyepiece holder and rebalances the telescope if necessary. The SSP-3 photometer weighs around three pounds and the SSP-5 weights around five pounds. The photometer must be turned on for at least 15 minutes outside before measurements commence to allow the electronics to stabilize and reach ambient temperature. One should also focus the telescope on a star and check to see that the focus does not change with time. Finally, one should allow the telescope to reach ambient temperature before making measurements.

There are three steps in making a brightness measurement, namely, (1) making readings of the sky, comparison star and target, (2) computing the target's magnitude and (3) computing the target's normalized magnitude and albedo.

While making measurements, one must be aware of their photometer's FOV, and must know how to aim his or her photometer to make the necessary measurements. Figure 5.9 shows the field-of-view (FOV) through the photometer. The large circle is the telescope FOV and the small one is the photometer FOV. Only light inside of the small circle will reach the detector. The SSP-3 and SSP-5 have small illuminated circles that show the photometer FOV. The size of the FOV depends on the photometer aperture and the telescope focal length. In some cases, the buyer can specify the photometer aperture. Optec Inc. can provide FOV information for their line of photometers. In addition, through the use of Barlow lenses and focal reducers, the observer can select the focal length for his/her telescope.

One can also determine the angular size of their photometer's FOV from the star test. Essentially one selects a star with a declination between 5°N and 5°S and then records how much time is required for this star to move across the FOV; the time in

Table 5.6. A summary of commercially available photoelectric photometers from Optec Inc., along with detectors, sensitivities, and other characteristics

Model	Detector	Faintest star [a] (V filter)	Wavelength range (nm) [b]
SSP-3	Photodiode	7.0	300 to 1100
SSP-4	Photodiode (InGaAs)	–	900 to 1870
SSP-5	Photomultiplier tube (R6350)	10	185 to 650
SSP-5	Photomultiplier tube (R6358)	13	185 to 830

[a] This is the faintest star that one can measure with ten 10-second integrations with an 11 inch telescope having a signal to noise ratio of 100.
[b] This refers to just the photometer. Telescope optics may decrease this range.

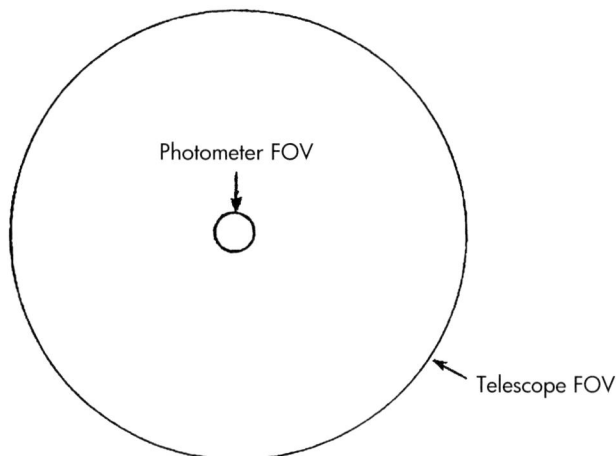

Figure 5.9. The large circle is the telescope field-of-view (FOV) and the smaller one is the photometer FOV. The photometer only measures light that passes through its FOV. (Credit: Richard W. Schmude, Jr.)

seconds is multiplied by (15 arc-sec/sec.) to get the size in arc-seconds. As an example, if it takes Delta-Orionis 12 seconds to move across a photometer's FOV, one computes an angular size of 12 sec. × (15 arc-sec/sec.) =180 arc-seconds.

The basic sequence in making a magnitude measurement is illustrated in Table 5.7 and in Figure 5.10. When one carries out these measurements, they should either write them down or have them sent to the computer. I record my data in a similar way to what is shown in Table 5.8. Two excellent books that describe the method are *Photoelectric Photometry of Variable Stars* by Douglas Hall and Russell Genet (Willmann-Bell Inc., 1988) and *Astronomical Photometry* by Arne Henden and Ronald Kaitchuck (Willmann-Bell Inc., 1990). In the next Section, I will show an example of how to analyze photoelectric magnitude data.

Example

In the next few pages, I will give a detailed example of a photometric run that I made on October 29, 2005. The measurements are in Table 5.8. All measurements were made near Barnesville, Georgia, which is at a latitude and longitude of 33.1°N, 84.1°W. The comparison star for all measurements was Sigma (σ)-Aquarii and the target was Uranus.

Table 5.7. Measurements needed in making a photoelectric photometry measurement; target = Uranus

Action	Look at
Three 10-second measurements of the sky near the comparison star	Figure 5.10A
Three 10-second measurements of the comparison star	Figure 5.10B
Three 10-second measurements of the sky near Uranus	Figure 5.10C
Three 10-second measurements of Uranus	Figure 5.10D
Repeat the above sequence two more times and then:	
Three 10-second measurements of the sky near the comparison star	Figure 5.10A
Three 10-second measurements of the comparison star	Figure 5.10B

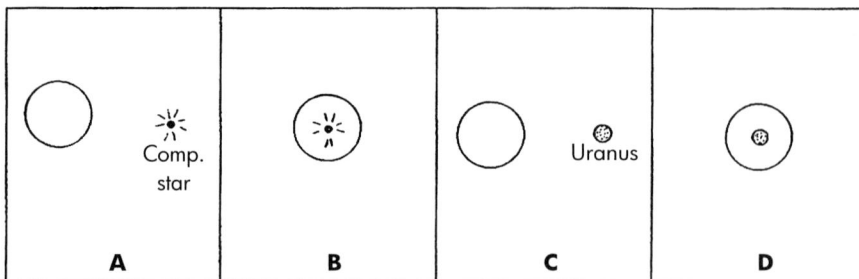

Figure 5.10. The basic sequence in making a brightness measurement. In all cases, the large circle is the photometer FOV. In frame A, one measures the sky brightness near the comparison star; in frame B, one measures the brightness of the comparison star; in frame C, one measures the sky brightness near Uranus; and in frame D, one measures the brightness of Uranus. (Credit: Richard W. Schmude, Jr.)

The first step is to compute the readings of the comparison star and the target. This is accomplished by subtracting the average sky brightness from the average sigma-Aquarii and Uranus readings. The resulting values are listed in the Diff. (or fifth) column in Table 5.8. The first Diff. value for Sigma-Aquarii is:

$$680.00 - 596.33 = 83.67$$

Since this is just a reading, it has no units. Next, I computed the Diff. values for the other readings. Keep in mind that the sky brightness readings contain both the dark current contribution along with scattered light form the sky and is to be made regardless of how dark the sky.

I computed the preliminary magnitude difference as:

$$\Delta v = -2.50 \times \log[U_{diff}/C_{diff}] \qquad (5.1)$$

where U_{diff} is the Diff. value for Uranus and C_{diff} is the average of the two adjacent Diff. values for Sigma-Aquarii. The first Δv value (sixth column) is:

$$\Delta v = -2.50 \times \log[30.00/83.33]$$
$$= -2.50 \times \log[0.360]$$
$$= -2.50 \times -0.4437$$
$$= 1.109$$

where C_{diff} in the above calculation equals $0.5 \times (83.6666 + 83.00) = 83.3333 \sim 83.33$.

What this says is that so far, Uranus is 1.109 magnitudes fainter than Sigma-Aquarii. The second and third Δv values are:

$$\text{Second } \Delta v \text{ (sixth column)} = -2.50 \times \log[30.33/81.33] = 1.071 \qquad (5.2)$$

$$\text{Third } \Delta v \text{ (sixth column)} = -2.50 \times \log[30.00/80.50] = 1.072 \qquad (5.3)$$

Let's take a time-out for some digression. I have just computed the magnitude differences between Uranus and Sigma-Aquarii. Furthermore, since three Δv values were computed, a more reliable Δv can be reported. There are two corrections that should be made before a final magnitude difference can be determined. These are: (1) color correction and (2) primary and secondary extinction corrections. Each correction is described below.

Table 5.8. Sample data recorded on Oct. 29, 2005

Time	Type	Readings	Avg. Read.	Diff.	Δv	AM	$k'_v \Delta AM$	$\varepsilon \Delta(B - V)$	ΔV
1:24	Sky	595, 597, 597	596.33						
1:25	σ-Aqr	680, 680, 680	680.00	83.67		1.388			
1:27	Sky	604, 603, 602	603.00						
1:28	Uranus	632, 633, 634	633.00	30.00	1.109	1.366	−0.006	−0.032	1.083
1:29	Sky	602, 606, 608	605.33						
1:29	σ-Aqr	687, 690, 688	688.33	83.00		1.387			
1:35	Sky	610, 607, 608	608.33						
1:36	Uranus	640, 640, 636	638.67	30.33	1.071	1.364	−0.006	−0.032	1.045
1:37	Sky	610, 615, 612	612.33						
1:38	σ-Aqr	692, 693, 691	692.00	79.67		1.386			
1:39	Sky	605, 607, 607	606.33						
1:40	Uranus	638, 634, 637	636.33	30.00	1.072	1.363	−0.006	−0.032	1.046
1:41	Sky	606, 603, 607	605.33						
1:42	σ-Aqr	687, 687, 686	686.67	81.33		1.386			

The time is listed in the first column, the object being measured is listed in the second column, the readings and average readings are listed in the third and fourth columns, the difference between the sky and object readings are listed in the fifth or Diff column, the preliminary magnitude difference is listed in the sixth column, the number of air masses that the measurement was made through is listed in the seventh (AM) column, the extinction and color correction terms are listed in the eighth and ninth columns, and the final magnitude difference is listed in the last column. In all cases, Sigma-Aquarii is abbreviated as σ-Aqr.

Different telescopes have different sensitivities to light. One Newtonian telescope with one type of mirror coating may be more sensitive to blue light than a second Newtonian telescope with a different coating. Furthermore, people use different photometers with different detectors having different color sensitivities. Because of these differences, several groups developed their own standard photometry system, which is a list of stars each having a magnitude assigned to it. One common system is the Johnson U, B, V, R and I System.

A color (or transformation) correction term should be measured for each filter used for each telescope-photometer combination. In my case, I use filters that are transformed to the Johnson B, V, R and I system. Each of these filters has a transformation coefficient. The Appendix describes how one can measure the transformation coefficient for the Johnson V System. For my SSP-3 photometer and Maksutov telescope, the transformation coefficients for the B, V, R and I filters are: 0.092, –0.051, –0.021 and –0.095 respectively. The color correction factor for the V filter is:

$$\varepsilon_v \Delta(B - V) = \varepsilon_v \times [(B - V)_{target} - (B - V)_{comp.star}] \tag{5.4}$$

where B–V is the color index.

Our atmosphere absorbs light. The amount of light absorbed depends on several factors including altitude of the target, wavelength, atmospheric conditions and how high the observer is above sea level. All of these factors are taken into account when extinction corrections are made.

Figure 5.11 shows a side view of a planetary atmosphere. Star B is at the observer's zenith, and so the light travels through one air mass of our atmosphere. The thickness of the atmosphere going straight up is defined as 1.0 air mass. Star A is at an altitude of 30° and as a result, light from it must travel through more air before reaching the observer; in fact, its light must travel through 2.0 air masses. Equation 5.5 relates the altitude of the star (A) in degrees and the approximate number of air masses (AM) that its light must travel through:

$$AM = 1/Sin(A) \tag{5.5}$$

This equation is satisfactory for altitudes greater than ∼10°. If the comparison star is at a different altitude than the target, an extinction correction should be made. The extinction correction for the V filter is:

$$\text{Extinction correction} = k_v' \times (AM_{target} - AM_{comp.star}) = k_v' \times \Delta AM \tag{5.6}$$

where k_v' is the extinction coefficient of the V filter, AM_{target} is the air mass of the target and $AM_{comp. star}$ is the air mass of the comparison star. The Appendix describes how one can evaluate k_v'. In order to determine the AM values, one must know the altitude A at the time of measurement. There are three ways of determining A, namely, (1) computer program, (2) measurement and (3) calculation. Methods 1 and 2 are much simpler.

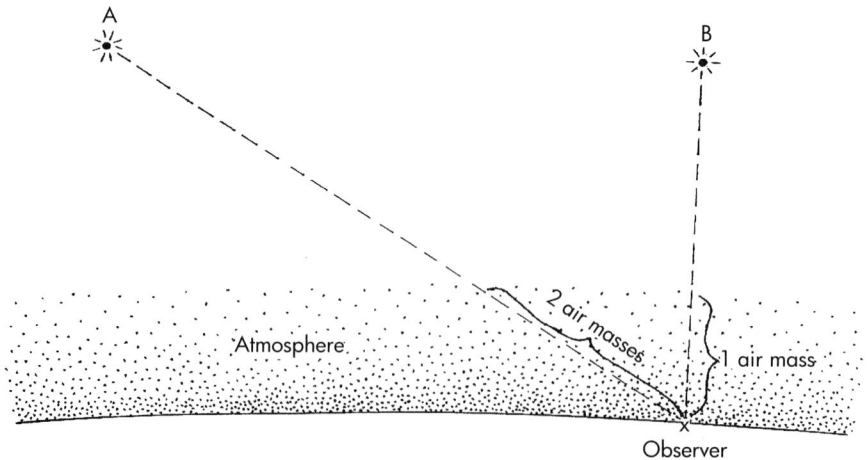

Figure 5.11. Side view of a planetary atmosphere. The thickness of an atmosphere going straight up is defined as one air mass. The lower the star elevation, the greater is the number of air masses that its light must travel through to reach the observer. (Credit: Richard W. Schmude, Jr.)

Several computer programs such as the JPL ephemerides generator will allow one to compute the altitude of any planet (as well as Pluto). One can also use a quadrant to measure the altitude of their comparison star or planet. The altitude can be measured to an accuracy of $1°$ or less. Figure 5.12 shows a quadrant and how one uses it. An example of how to calculate the altitude of a celestial object is given in the Appendix.

For the ultraviolet and blue filters (Johnson system) one must also make an additional correction, which is:

$$k'' \times AM_{avg} \times [(B - V)_{target} - (B - V)_{comp.star}] = k'' \times AM_{avg} \times \Delta(B - V) \quad (5.7)$$

where k'' is the color-dependent extinction coefficient, AM_{avg} is the average air mass of the target and comparison star and the other terms are the same as before. For the U and B filters, one can assume that $k'' = -0.03$ and for the V, R and I filters, k'' can be assumed to equal 0. If one desires an accuracy of 0.001 magnitudes then one should measure k'' for their system.

Returning now to Table 5.8, I have computed the corrected magnitude difference as:

$$\Delta V = \Delta v - [k'_v \times (\Delta AM)] - [k'' \times AM_{avg} \times \Delta(B - V)] + [\varepsilon_v \times \Delta(B - V)] \quad (5.8)$$

where $k'_v = 0.26$ magnitude/air mass, $k'' = 0.00$ and $\varepsilon_v = -0.051$. The first ΔV value for Uranus in Table 5.8 is:

$$\Delta V = 1.109\,(\text{sixth column}) - [0.26 \times (1.366 - 1.3875)] - [0] + [-0.051] \times (0.56 - -0.06)$$

$$= 1.109 - -0.006 - 0 - 0.032\,(\text{sixth, eighth and ninth columns})$$

$$= 1.083\,(\text{tenth column})$$

The other two ΔV values are computed in the same manner.

Table 5.9 lists the ΔV values along with measured and normalized magnitudes of Uranus. Magnitudes of Uranus (V) are computed as:

$$V = \Delta V + \text{magnitude of the comparison star} \quad (5.9)$$

Since the V filter magnitude of Sigma-Aquarii is 4.82, the first V filter magnitude is:

Figure 5.12. A quadrant has a weight, suspended on a string that points straight down. When a quadrant is lined up on a target, one can determine the target's altitude by measuring where the weight is in relation to the protractor. (Credit: Timothy Abbott and Richard W. Schmude, Jr.)

Table 5.9. Measured and normalized magnitudes of Uranus computed from data collected on Oct. 29, 2005

Time	ΔV	Measured Magnitude	Uranus-Earth Distance (au)	Uranus-Sun Distance (au)	V(1,0)	Vo
1:28	1.083	5.903	19.539	20.072	–7.06	5.65
1:36	1.045	5.865	19.539	20.072	–7.10	5.61
1:40	1.046	5.866	19.539	20.072	–7.10	5.61

The universal Time is listed in the first column, followed by the magnitude difference, measured magnitude, Uranus-Earth, and Uranus-Sun distances. The normalized magnitude reduced to a distance of 1.0 au is listed in the sixth column and the normalized magnitude reduced to a mean opposition is listed in the seventh column.

$$V = 1.083 + 4.82$$

$$= 5.903 \sim 5.90.$$

The other two V filter magnitudes are computed in a similar manner. The V filter magnitudes are rounded off to the nearest 0.01 magnitudes since the comparison star magnitudes are reported to this amount and the uncertainty of the measurement is $\sim$0.02 magnitudes.

One problem with the V filter magnitude is that it changes with distance. As Uranus gets farther from us, its magnitude changes. The normalized magnitude, however, does not change with changing target distance. There are two ways of calculating a normalized magnitude. Each of these is described.

One way of computing the normalized magnitude, V(1,0), is to compute the magnitude of the target if it is 1.0 au from both the Earth and the Sun and if its solar phase angle equals $0°$. This is accomplished as:

$$V(1,0) = V - 5.00 \times \log[r \times d] + 2.5 \times \log[k] - (c_v \times \alpha). \qquad (5.10)$$

where r is the Uranus-Sun distance, d is the Uranus-Earth distance, k is the fraction of Uranus's disc that is illuminated, c_v is the solar phase angle coefficient of Uranus and α is the solar phase angle of Uranus. Values of r and d are listed in Table 5.9 and should be in astronomical units (au). The value of k will always exceed 0.999 for Uranus and, hence, the $2.5 \times \log [k]$ term is negligible. The $c_v \times \alpha$ term is also negligible because α never exceeds $3°$ for Uranus and c_v is very small. The first V(1,0) value in Table 5.9, sixth column is:

$$V(1,0) = 5.903 - 5.00 \times \log[20.072 \times 19.539]$$

$$= 5.903 - 5.00 \times \log[392.1868]$$

$$= 5.903 - 5.00 \times 2.5935$$

$$= 5.903 - 12.967$$

$$= -7.064 \sim -7.06$$

The other two V(1,0) values are computed in the same way.

One can also compute the magnitude at average opposition, Vo. This is the magnitude that a planet has when it is at opposition and is at its average distance from the Sun. The Vo value for Uranus is computed as:

$$Vo = V - 5.00 \times \log[(r/19.191\,au) \times (d/18.191\,au)] \qquad (5.11)$$

Where the r and d values are defined in equation 5.10, the 19.191 au is the average Uranus-Sun distance and the 18.191 au is the average Uranus-Earth distance at opposition. As an example, the first Vo value in Table 5.9, seventh column, is:

$$Vo = 5.903 - 5.00 \times \log[(20.072\,au/19.191\,au) \times (19.539\,au/18.191\,au)]$$

$$= 5.903 - 5.00 \times \log[1.0459 \times 1.0741]$$

$$= 5.903 - 5.00 \times \log[1.1234]$$

$$= 5.903 - 5.00 \times 0.05054$$

$$= 5.903 - 0.2527$$

$$= 5.650 \sim 5.65$$

The other Vo values are computed in the same way.

As mentioned in an earlier Section, brightness measurements can yield information on changes in a planet's color and albedo. I would like to describe how one looks for these changes.

One can measure color changes by measuring the brightness of the target in different color filters. One way of doing this is to measure the target's brightness in filters transformed to the Johnson B, V, R and I system and then compute the color index. The color index is the difference in magnitude between two different filters. For example, Neptune's B–V color index is around 0.4 and so the B filter magnitude minus the V filter magnitude equals 0.4 magnitudes or simply 0.4. If a planet's color changes, then its B–V value will change.

One can also measure albedo changes taking place on a planet by making brightness measurements. Before one can measure these changes, he or she should know how to compute a planet's albedo. There are several ways of defining the albedo of a planet. One form of albedo is the geometric albedo, which was defined in chapter one. The geometric albedo (p) of a solar system object is computed as:

$$p = \text{inverse } \log[0.4 \times \{V(1,0)_{Sun} - V(1,0)_t\} - 2\log(\sin(\sigma.))] \qquad (5.12)$$

where $V(1,0)_{Sun}$ is the normalized magnitude of the Sun, $V(1,0)_t$ is the normalized magnitude of the target and σ is the angular size of the radius of the target at a distance of 1.0 astronomical unit in degrees. The most recently measured normalized magnitudes of the Sun are: V = –26.75, B = –26.10, R = –27.27 and I = –27.63. Pluto probably has no polar flattening since it rotates slowly and, hence, its σ value is 0.000444° whereas Pluto + Charon has an effective value of σ = 0.000500°. Since Uranus and Neptune have a small amount of polar flattening, one should compute the value of σ from both the polar and equatorial radii. I computed σ values for several sub-Earth latitudes (s) for both planets and then fit the σ and s values to a quadratic equation with a least squares routine. The resulting equations are 5.13

and 5.14. One can use them to compute values of σ for Uranus and Neptune at any sub-Earth latitude.

$$\sigma = 0.009671 + (1.087 \times 10^{-6})\,s + (5.184 \times 10^{-9})\,s^2 \text{ [Uranus]} \quad (5.13)$$

$$\sigma = 0.009403 - (6.814 \times 10^{-8})\,s + (2.602 \times 10^{-8})s^2 \text{ [Neptune]} \quad (5.14)$$

I will compute the geometric albedo of Uranus based on the first V(1,0) value measured on Oct. 29, 2005 (per Table 5.9):

$$p = \text{inverse log } [0.4 \times \{-26.75 - -7.06\} - 2\log(\sin(0.00968))]$$

$$p = \text{inverse log } [0.4 \times \{-19.69\} - 2\log(0.000169)]$$

$$p = \text{inverse log } [-7.876 + 7.544]$$

$$p = \text{inverse log } [-0.332]$$

$$p = 0.466 \text{ or }.47$$

One can use photometry to measure a planet's brightness change as it rotates; or, in other words, its diurnal brightness change. A graph of brightness versus a planet's longitude is called a rotational light curve. Diurnal changes occur when a planet rotates. The easiest way to measure diurnal changes for Uranus and Neptune is to make several brightness measurements over a six to eight hour period on each of two consecutive nights. This data set will show a complete rotational light curve. This is possible because we see the opposite side of these planets on consecutive nights. The reason for this is that these planets have rotation periods equal to about two-thirds of an Earth day. The situation is different for Pluto. One can measure diurnal changes by making one brightness measurement per night for at least a week. This is because of Pluto's longer rotation period.

A useful project would be to measure diurnal brightness changes of Uranus and Neptune using near-infrared light. Thin, high-altitude hazes on Uranus and Neptune often reflect a much greater percentage of this light than the rest of the planet. As a result, when these hazes come into view, they can cause a brightness surge of 10% or more. Once they rotate out of view, the planet returns to its normal brightness.

CCD Photometry

One can use a CCD camera along with appropriate filters and software to measure the brightness of Uranus, Neptune, and Pluto. There are several points to remember when doing CCD photometry (1) one should make flat field and dark frame corrections to all images that will be used for photometry; (2) one should measure transformation coefficients for their camera-filter-telescope system; (3) one can average several images, but must not do other image processing such as contrast enhancement; (4) one should not overexpose images; (5) exposure times should be long enough to prevent a shutter induced brightness gradient; (6) the comparison star and target should be in the same image; and (7) for B filter measurements, a secondary extinction correction should be made. Points 1-4 and 6-7 need no further explanation.

As for Point 5, a shutter will block some parts of the image before other parts. This can lead to a systematic error. If the exposure time exceeds ~1 second then this should not be a problem. Doug West has collected V filter brightness data of Pluto with a CCD camera; his technique is summarized in Table 5.10. Essentially, Doug subtracted the sky brightness from his comparison star and Pluto reading before analysis. His Pluto results are listed in Table 5.11. The normalization and color corrections were made in the same way as discussed in the photoelectric photometry Section above.

Doug also used his CCD camera to measure the brightness of Uranus and Neptune. His technique was similar to what was used for Pluto. His results were consistent with those measured with the SSP-3 photometer.

Sources of Error in Photometric Measurements

Ten sources of error in photoelectric magnitude measurements are (1) random error, (2) extinction error, (3) seeing, (4) small photometer FOV, (5) variable sky transparency, (6) transformation error, (7) position of object in photometer FOV, (8) temperature, (9) variable comparison star and (10) stars too close to the target or to the comparison star. I will talk about how to reduce errors from each of these sources.

Random error is caused by noise along with small changes in the electronics of the photometer or CCD camera. One can reduce random error by either using a larger telescope, using a more sensitive detector or by taking additional measurements. One can also reduce random error by using as small a photometer aperture as possible. This is especially important when carrying out brightness measurements of moons. A large aperture allows too much scattered light to reach the detector, which is a source of random error. If one chooses to carry out brightness measurements of objects on digital images he or she may be able to select the appropriate aperture from their software package. In many cases, the operator can experiment with different apertures until he or she obtains the optimum result.

Table 5.10. Summary of the technique that Doug West used in measuring the brightness of Pluto using his CCD camera

Camera used	ST-8 and ST-9E made by SBIG
Filter used	Transformed to Johnson V system
Telescope characteristics	(0.2 m) 8 inch Schmidt-Cassegrain; 50 inch focal length
Camera-computer connection	Parallel port
Exposures used	Five 30-second exposures were averaged
Dark current correction	Yes
Flat field correction	Yes
Software	Mira software was used in averaging the images and extracting the photometric data. SBIG software was also used.
Comparison star source	Tycho star catalog
Aperture size (photometry)	12 arc-seconds
Pixel size in images	3.4 arc-seconds
Computer used	PC with a parallel port; windows operating system

Table 5.11. Summary of Doug West's Pluto + Charon V filter photometric measurements

Date	Measured magnitude [a]	Solar Phase Angle	V(1,0)	Geometric Albedo
Apr. 17.419, 2001	13.63	1.4°	−1.20	0.94
Apr. 23.429	14.06	1.2°	−0.76	0.63
Apr. 25.423	14.12	1.2°	−0.70	0.60
May 1.423	13.79	1.1°	−1.02	0.80
May 7.426	13.75	0.9°	−1.05	0.82
May 8.425	14.14	0.9°	−0.66	0.57
May 14.425	13.62	0.7°	−1.17	0.92
May 19.170	13.83	0.6°	−0.95	0.75
May 24.177	13.88	0.6°	−0.90	0.72
Apr. 28.450, 2002	13.86	1.2°	−0.98	0.77
Mar. 8.480, 2003	14.14	1.9°	−0.80	0.65
Mar. 10.476, 2003	14.04	1.9°	−0.90	0.72
Mar. 12.462	13.77	1.9°	−1.17	0.92
Mar. 15.471	13.72	1.9°	−1.21	0.95
Mar. 21.485	13.08	1.8°	−1.84	–
Mar. 26.463	13.86	1.8°	−1.06	0.83
Mar. 17, 2004	14.14	1.8°	−0.81	0.66
Mar. 19.463	13.87	1.8°	−1.08	0.85
Mar. 20.417	14.05	1.8°	−0.89	0.71

[a] Estimated uncertainties for the magnitudes are 0.05 to 0.10 magnitudes.

One can reduce errors caused from atmospheric extinction by measuring accurately the extinction coefficient. Another method is to carry out measurements when the target and comparison star are at high altitudes. The greater the altitude of the comparison star and target, the less light that is lost due to absorption by the Earth's atmosphere. Finally, extinction error may be reduced by carrying out measurements when the sky is as transparent as possible. Skies are often hazy during the summer months in the United States. Therefore, I make very few brightness measurements during this time. I have found that skies are the most transparent right after a cold front has passed.

One should carry out measurements under average or better seeing conditions because the steadier the air, the more likely one is able to collect all of the target's photons when it is in the photometer FOV. See Figure 5.13. Under poor seeing conditions, the target (or comparison star) appears to move around in the FOV, resulting in the loss of some light. See Figure 5.14.

A small photometer FOV (less than 10 arc-seconds) will not collect all of the light from the target and comparison star as is shown in Figure 5.15. For most situations, a photometer FOV of around 40 arc-seconds for a 0.2 m (8 inch) telescope is satisfactory. A smaller FOV can be used for larger telescopes with good mounts.

A changing sky transparency will lead to error. One way to detect this change is to carry out several measurements as discussed earlier in this Chapter. If the photometer readings jump around by more than a couple of percent, this would be a sign of variable transparency. If one is using a traditional single channel

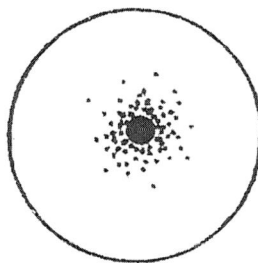

Figure 5.13. When the seeing is good, almost all of the light from the target lies within a few arc-seconds of the target's center. As a result, almost all of the light enters the photometer FOV. The dots represent scattered light and the circle represents the photometer FOV. (Credit: Richard W. Schmude, Jr.)

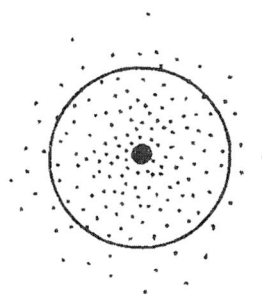

Figure 5.14. Under poor seeing, some light from the target never reaches the photometer FOV. This in turn affects the measured brightness. The dots represent scattered light and the circle represents the photometer FOV. (Credit: Richard W. Schmude, Jr.)

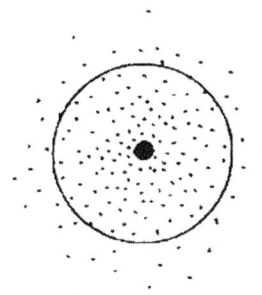

Figure 5.15. A small photometer FOV does not capture all of the light from the target, thereby affecting the measurement. The dots represent scattered light and the circle represents the photometer FOV. (Credit: Richard W. Schmude, Jr.)

photometer, like the SSP-3 instrument, they should cease measurements when the transparency changes. If, however, one is using a CCD camera, variable transparency will not be as serious. This is because the transparency is usually the same for both the comparison star and target.

Any error in the transformation coefficients will show up in the final results. One way to reduce transformation errors is to measure the transformation coefficients on a clear night with good seeing. One should make several measurements of their transformation coefficients and determine average values. In the best case, one can estimate the uncertainties in his or her transformation coefficients, and, from this, estimate the error due to transformation. A second way to reduce transformation related errors is to select a comparison star having a color as close to that of the target as possible.

Another source of error is inconsistent placement of the target or comparison star in the photometer FOV. One should place the target and comparison star in the same location within the photometer FOV. I have noticed that when a star is

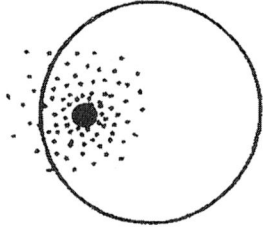

Figure 5.16. When the target is not centered in the photometer FOV, some of the light may not reach the detector. This will lead to an incorrect measurement. The dots represent scattered light and the circle represents the photometer FOV. (Credit: Richard W. Schmude, Jr.)

centered in the FOV the reading is 1–2% higher than when it is halfway between the center and the edge of the field. See Figure 5.16.

Another source of error is temperature. The temperature can change the wavelength of the peak transmission of a filter and its FWHT. To reduce temperature errors, one should carry out the transformation measurements at a temperature that is as close to the measurements as possible.

A variable comparison star – the ninth source of error above – should not be used unless it varies by less than the uncertainty of the measurement. One can check for variability by using one or more check stars. Before using a new comparison star, one should also check to see if it is a variable star. In some cases, a double star can also pose a problem. If the separation is greater than a few arc-seconds, one should decide whether to use one or both stars.

A final source of error is a bright star near the target or comparison star. One will usually be able to see a star near Uranus or Neptune that is bright enough to affect the readings. This has stopped me from making measurements on several occasions. The situation for Pluto is different. A magnitude 15.5 star near Pluto will increase the brightness by over 20% and yet it would be very hard to see such a star. One can reduce their chances of this kind of error by using a CCD camera instead of a photoelectric photometer when doing magnitude measurements for Pluto. The CCD camera is more sensitive to faint stars than the eye and telescope.

In some cases, one must make compromises to reduce such errors. On many occasions, for example, I have chosen to make measurements on nights right after a cold front has passed. During these times, the atmosphere was usually nearly haze-free, but the seeing was usually poor. Errors due to poor transparency are usually more serious than those caused by poor seeing.

Spectra

How can astronomers determine the chemical composition of a planet's atmosphere? The astronomer first measures the intensity of different wavelengths of light. This is called a spectrum. The astronomer measures a planet's spectrum with a telescope, a spectroscope and a camera. Once the spectrum is available, the astronomer compares it to the spectra of compounds like methane and ammonia. If, for example, the methane spectrum matches that of Neptune, then one can conclude that methane is in Neptune's atmosphere.

Frank Melillo, the New York amateur astronomer mentioned earlier, imaged the spectrum of Uranus. The raw, black-and-white spectrum is shown in Figure 5.17. Melillo then converted the spectrum into a graph of light intensity (vertical axis) versus the wavelength of light (horizontal axis). The graph is shown in Figure 5.18.

Figure 5.17. Spectrum of Uranus imaged by Frank Melillo using a 0.25 m (10 inch) Schmidt-Cassegrain f/10 telescope on Sept. 12, 2006. (Credit: Frank Melillo, Holtsville, NY, ALPO Remote Planets Section.)

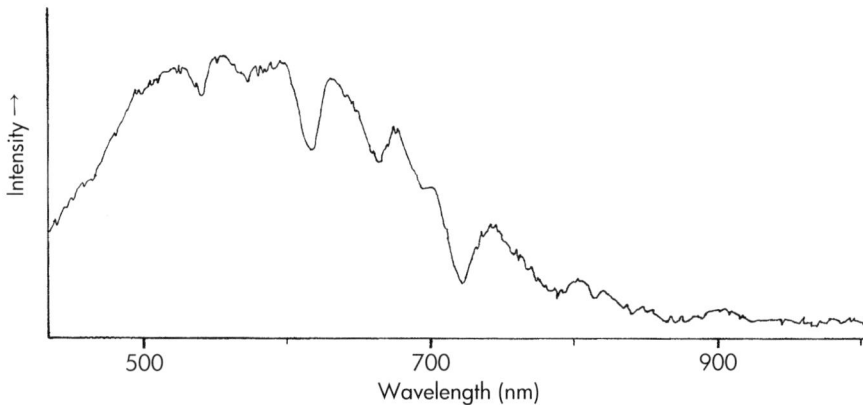

Figure 5.18. Spectrum of Uranus converted to an intensity versus wavelength plot, based on Frank Melillo's Sept. 12, 2006 Uranus spectrum. (Credit: Frank Melillo, Holtsville, NY, ALPO Remote Planets Section.)

Figures 5.17 and 5.18 are each called a spectrum. Table 5.12 summarizes Melillo's method for obtaining Uranus's spectrum.

A record of planetary spectra over several years will show changes in color and chemical composition. This kind of data can enable one to monitor long-term atmospheric changes. Furthermore, one can assess differences between the polar and equatorial regions. A long record of spectra may also explain the seasonal color change observed for Uranus between 1986 and 2006. See Chapter 1. Due to the long seasons on the remote planets, spectra over long periods of time are needed in order to understand better seasonal changes.

Visual Magnitude Estimates of Pluto

There are two exciting projects in which people making visual magnitude estimates of Pluto can participate. These are measuring Pluto's light curve and determining if Pluto's atmosphere is freezing out. Pluto rotates once every 6.387 days, causing the bright and dark sides of it to face Earth. As a result, Pluto's brightness changes. Thus one can plot Pluto's brightness versus time which is a

Table 5.12. Summary of the method that Frank Melillo used in obtaining the spectrum of Uranus on September 12, 2006

Telescope	10 inch Schmidt-Cassegrain f/10
Camera	Starlight Xpress MX-5
Arrangement of equipment	CCD adapter hooked to the telescope; Transmission grating spectroscope attached to the adapter and CCD camera inserted at the end
Barlow lens	No
Software used	Starlite Xpress; Photoshop Pro 8
Exposure time	10 seconds
Spectrum Range	400 to 900 nm; 500 pixels spanned the spectrum
Technique	Starlight Xpress software was used to get the intensity profile. One row of pixels was then fed into Photoshop pro 8 software to stretch the spectrum into an intensity versus wavelength graph.

light curve. During the 1990 s, Pluto's brightness changed by 0.3 magnitudes as it rotated. I believe that visual observers will be able to detect this change.

A second project involves carrying out visual magnitude estimates of Pluto over several years. One can then compute Pluto's albedo over a several year period and look for changes. If Pluto's albedo increases it would be consistent with the atmosphere freezing out onto its surface.

One can estimate Pluto's visual magnitude in the same way that is described in Chapter 4. The path of Pluto from 2008 through 2020 is presented in Figure 5.19. Since Pluto is near magnitude 13.7, faint comparison stars are needed. I have reproduced AAVSO charts of six variable stars near Pluto (UZ-Serpentis, HS-, SS-, FN-, TW- and Z-Sagittarii), which contain the magnitudes of faint comparison

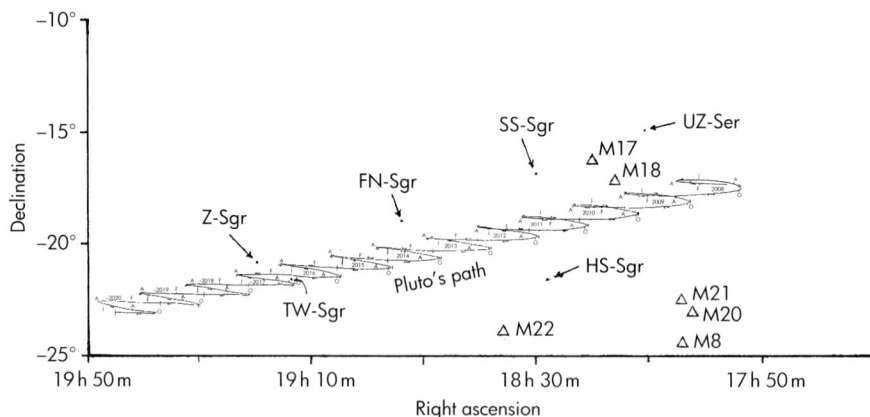

Figure 5.19. The path that Pluto will follow from Jan. 1, 2008 to Jan. 1, 2021. The tiny letters F, A, J, A and O for each year are the locations that Pluto will be on the first of February, April, June, August and October respectively. The tiny vertical line at the beginning of each year represents the position of Pluto on Jan. 1. The positions of stars with suitable comparison stars for visual magnitude studies of Pluto are also shown. (Credit: Richard Schmude, Jr., the American Association of Variable Star Observers and the Jet Propulsion Laboratory.)

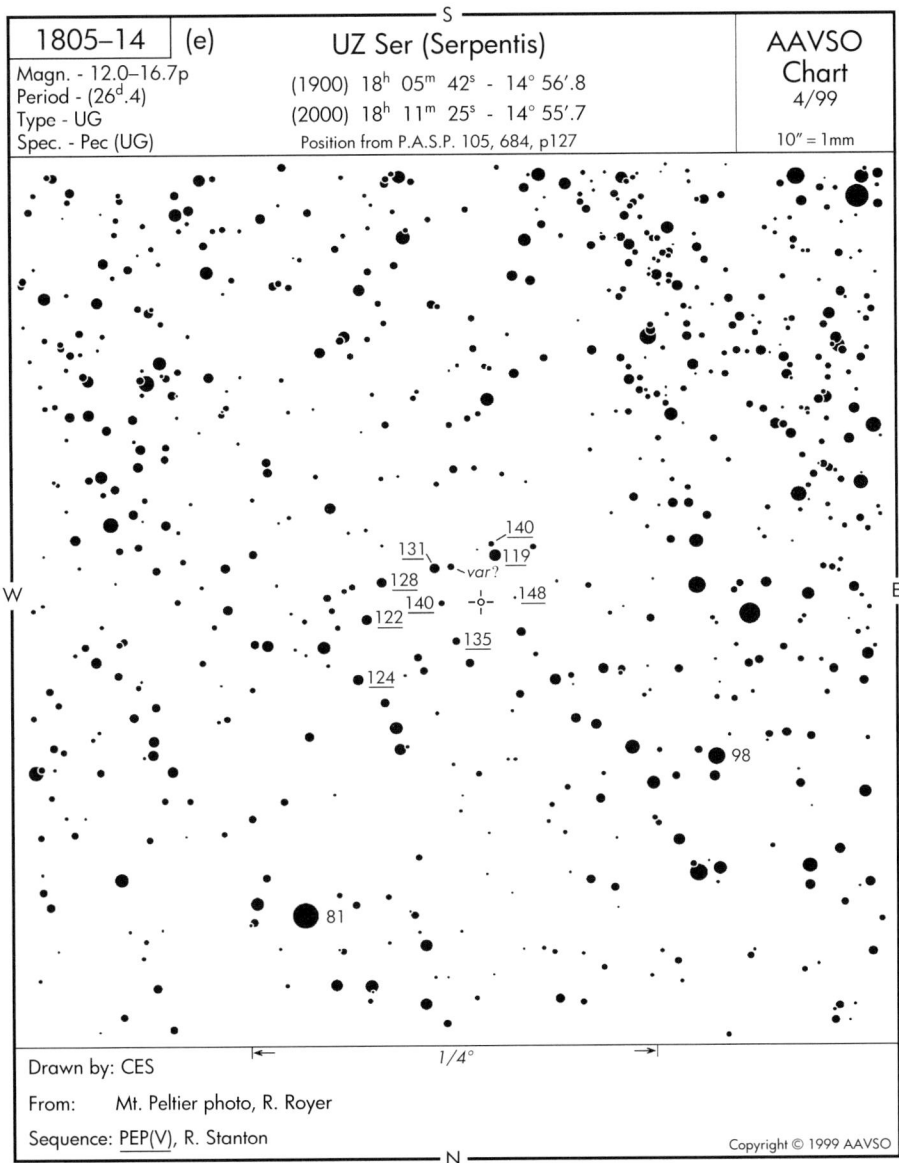

Figure 5.20. Finder chart and comparison stars for UZ-Serpentis. The comparison star magnitudes lack a decimal point; therefore, a star that has 140 next to it has a visual magnitude of 14.0. The comparison stars are suitable for estimating visual magnitudes of Pluto. (Credit: the American Association of Variable Star Observers, Cambridge, MA.)

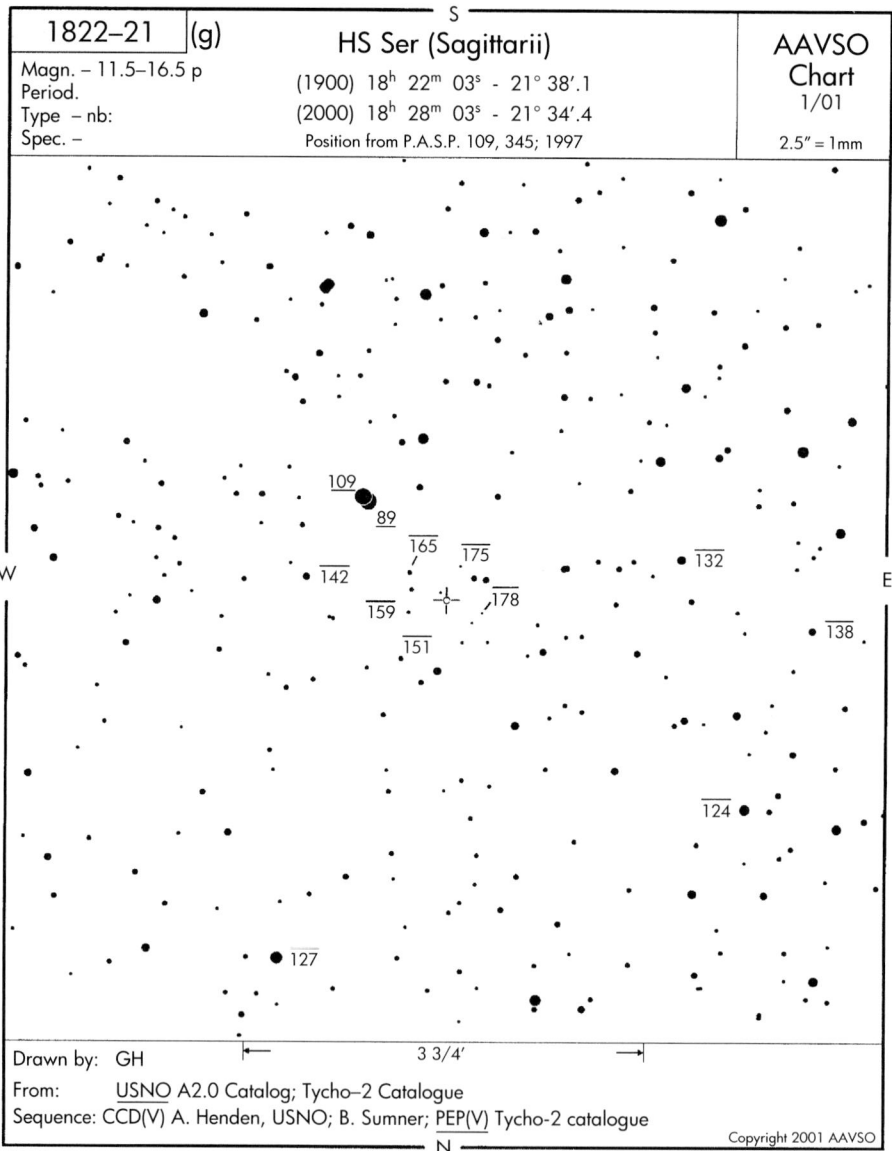

Figure 5.21. Finder chart and comparison stars for HS-Sagittarii. (Credit: the American Association of Variable Star Observers, Cambridge, MA.)

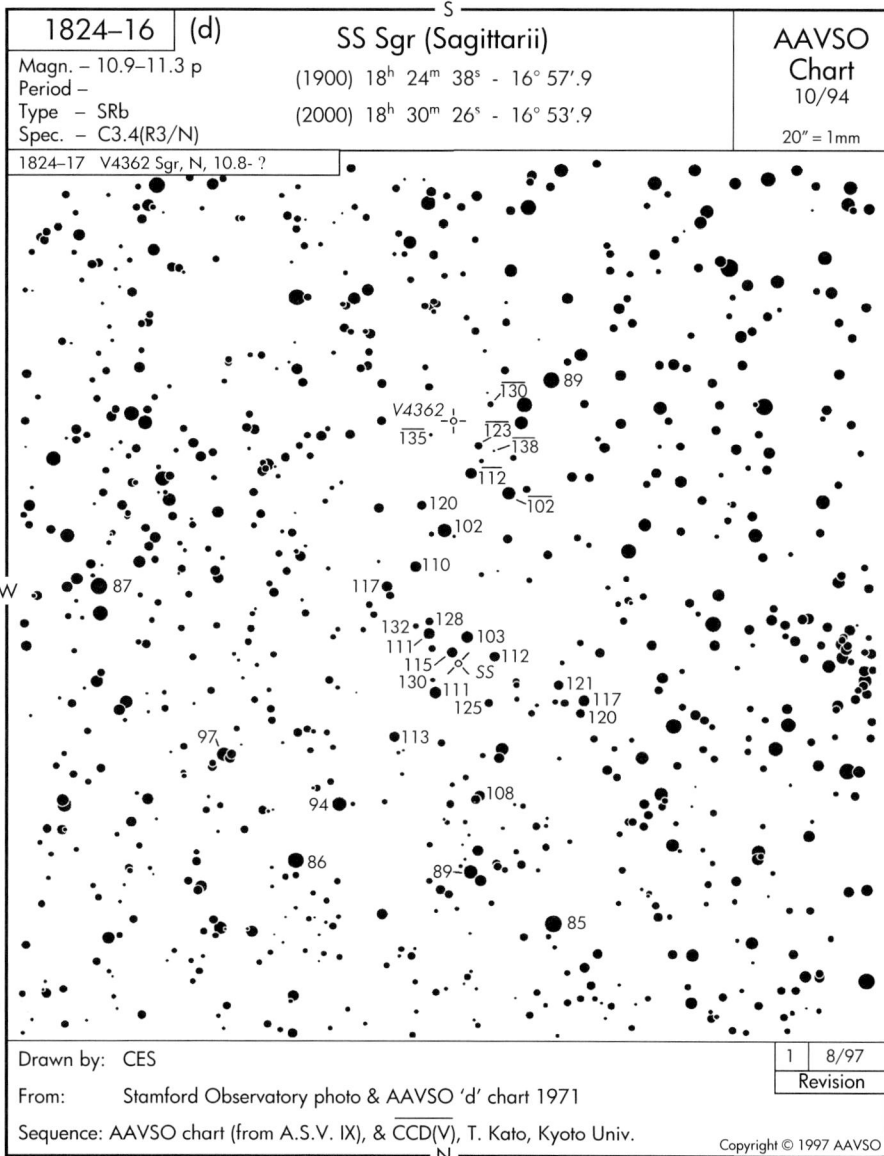

Figure 5.22. Finder chart and comparison stars for SS-Sagittarii. (Credit: the American Association of Variable Star Observers, Cambridge, MA.)

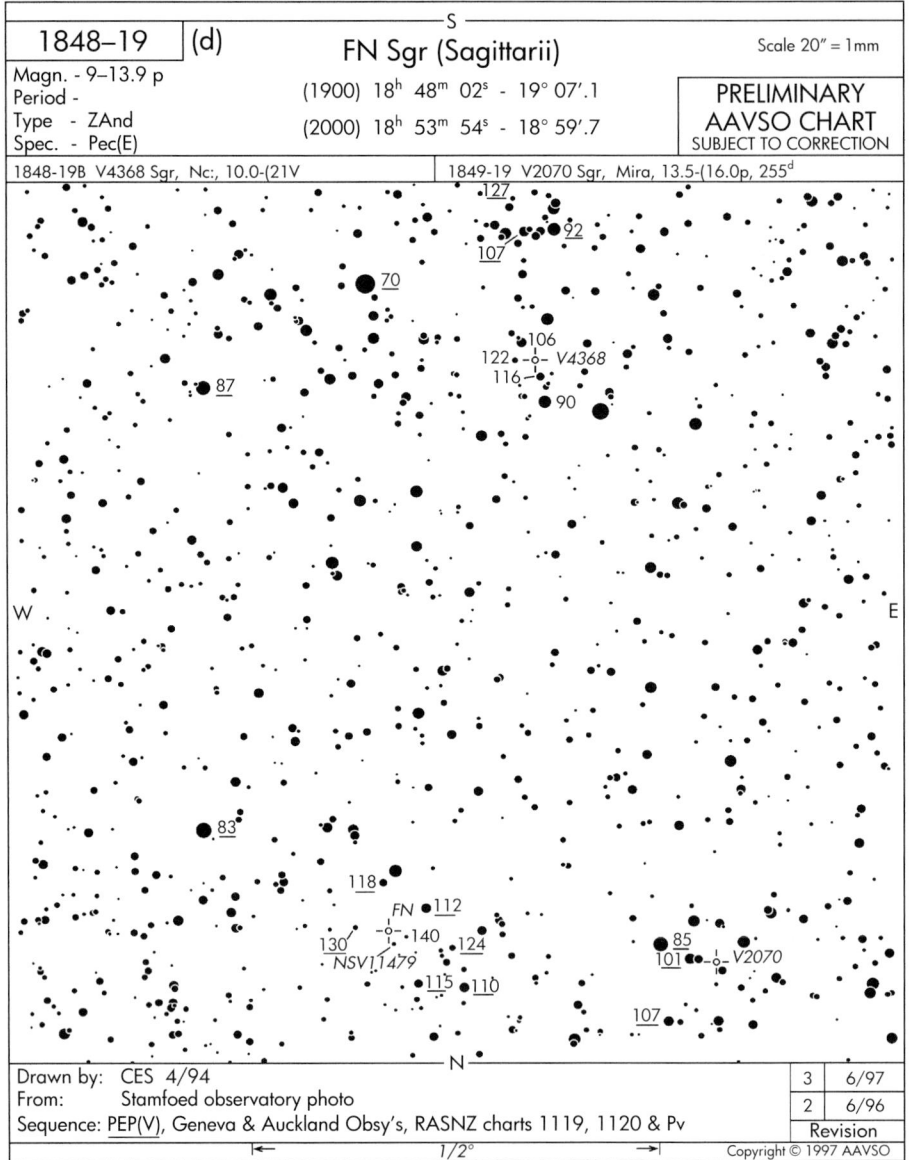

Figure 5.23. Finder chart and comparison stars for FN-Sagittarii. (Credit: the American Association of Variable Star Observers, Cambridge, MA.)

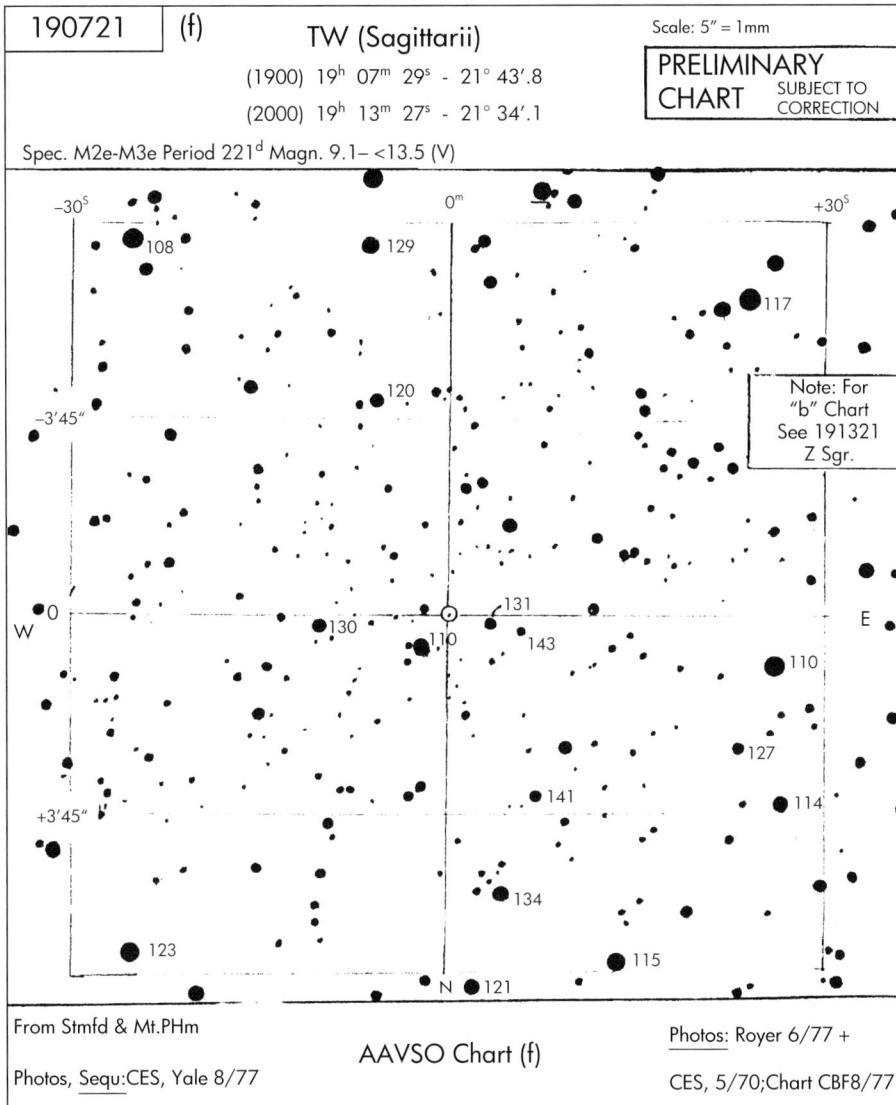

Figure 5.24. Finder chart and comparison stars for TW-Sagittarii. (Credit: the American Association of Variable Star Observers, Cambridge, MA.)

stars. These charts are reproduced in Figures 5.20–5.25. One estimates Pluto's brightness in terms of the faint stars near one or more of these variables. One will have to move the telescope between Pluto and the comparison stars. An automated telescope should work especially well. One must remember that the last digit in the star magnitudes on the AAVSO charts is the tenths decimal place, no decimal is included since it can be mistaken for a star. As an example, the star just below the center of UZ-Serpentis is labeled as 135 but the magnitude of this star is 13.5.

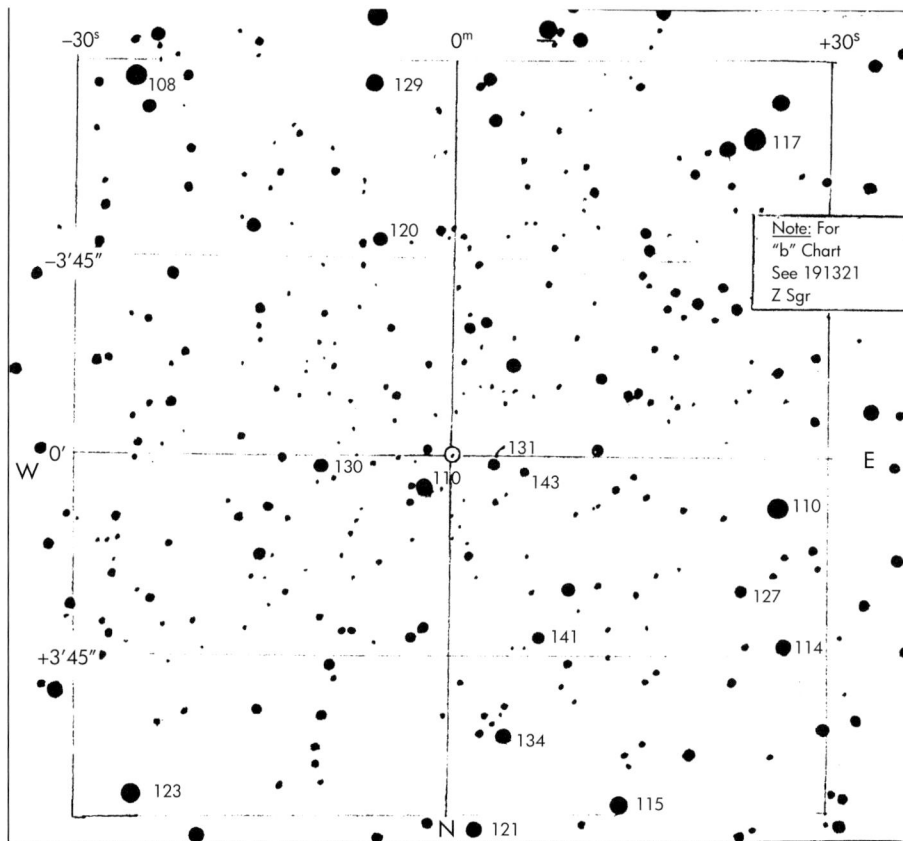

Figure 5.25. Finder chart and comparison stars for Z-Sagittarii. (Credit: the American Association of Variable Star Observers, Cambridge, MA.)

Table 5.13. Coordinates for the variable stars and associated comparison stars that can be used for making visual magnitude estimates of Pluto

Star	Field width (degrees)	Right Ascension (2000.0)	Declination (2000.0)	Right Ascension (2020.0)	Declination (2020.0)
UZ-Ser	0.54	18 h 11.4 m	−14° 55.7′	18 h 10.3 m	−14° 55.5′
HS-Sgr	0.14	18 h 28.1 m	−21° 34.4′	18 h 26.9 m	−21° 33.7′
SS-Sgr	1.1	18 h 30.4 m	−16° 53.9′	18 h 31.6 m	−16° 53.1′
FN-Sgr	1.1	18 h 53.9 m	−18° 59.7′	18 h 55.1 m	−18° 58.2′
TW-Sgr	0.23	19 h 13.5 m	−21° 34.1′	19 h 14.6 m	−21° 32.2′
Z-Sgr	0.35	19 h 19.7 m	−20° 56.0′	19 h 20.9 m	−20° 53.9′

The coordinates of the variable stars are listed in Table 5.13 along with the angular sizes of the chart widths (or field widths) in Figures 5.20–5.25. The right ascension and declination values of stars change as a result of the precession of Earth's axis. Between 2000 and 2020, the right ascensions of the variable stars will shift by about 15–20 arc-minutes. As a result, I have listed coordinates for the years 2000 and 2020. To find the exact coordinates for a particular year, one should interpolate between the related years.

Chapter 6

Observing with Large Telescopes

There are several projects involving remote planets that one can do with telescopes having diameters larger than 0.25 m (10 inches). These generally require more light than those described in the previous two chapters. A few of these projects include making drawings and images, measuring ellipticity values, making methane band images, carrying out methane band photometry, making polarization measurements, and timing satellite eclipses and transits. Although few amateurs have telescopes with diameters over 0.5 m (20 inches), there are several astronomy clubs with large telescopes, including the Atlanta Astronomy Club (Georgia), the Fort Bend Astronomy Club (near Houston, Texas) and the Salt Lake City Astronomy Club (Utah). These clubs have their own rules about who can use the telescopes but, in most cases, only club members with some training are allowed to use these instruments. With these large telescopes, one can carry out important work on remote planets.

Before various projects are discussed, let's discuss the orientation of Uranus and Neptune. This is critical for determining the location of albedo features, measuring the ellipticity, and identifying satellites.

Orientation

The orientation shows the north (N), south (S), east (E), and west (W) directions for either the observer's sky or the planet being studied. Unlike Jupiter and Saturn, it is difficult to determine the north-south orientation of Uranus and Neptune. This is because there is almost no detail on these planets; furthermore, Uranus and Neptune have very little polar flattening compared to Jupiter and Saturn. One can describe orientation as the observer's N, S, E and W sky directions or as the planet's N, S, E and W directions. I will describe how to determine both types of orientation. It is essential that the observer state which type of orientation (sky or planet) applies.

To find the N and S sky directions, one nudges the telescope in the north direction (towards the North Star) and the planet will move in the south sky direction. See Figure 6.1. One can also nudge the telescope in the south direction and the planet will move in the north sky direction. If one wants to know the approximate west sky direction, he/she needs to cut off the telescope drive and watch the planet's disc move across the FOV. The preceding limb is close to the west sky direction. This is shown in Figure 6.2.

Another way to determine the orientation of the sky is to note the position of star patterns near the planet and compare them to star patterns in a sky atlas. From this, one can determine the directions of N, S, E and W in their sky. Binoculars can be a great help in establishing orientation. Keep in mind that these directions are for the observer's sky and not the target.

R.W. Schmude, Jr., *Uranus, Neptune, and Pluto and How to Observe Them*,
DOI: 10.1007/978-0-387-76602-7_6, © Springer Science+Business Media, LLC 2008

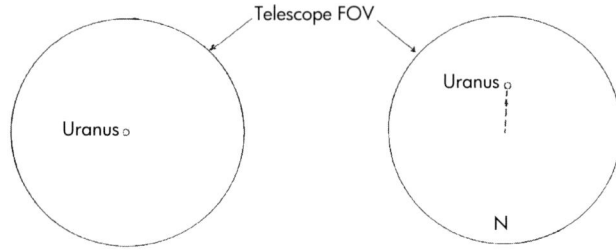

Figure 6.1. When one nudges the telescope in the north direction, the southern part of the sky can be identified because it is into which the planet appears to move. (Credit: Richard W. Schmude, Jr.)

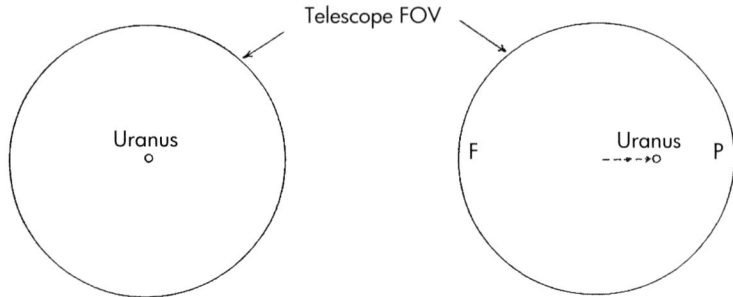

Figure 6.2. When one turns off the clock drive, the planet will move in the west direction in the sky. The leading edge of the planet is the preceding side and is labeled as p. (Credit: Richard W. Schmude, Jr.)

Determining Uranus's north, south, east and west limbs is more difficult because of the tilt of that planet's axis, and the fact that its axis generally does not lie in the same plane as Earths axis. The same problem exists for Neptune. The easiest way to get the planet's orientation is to determine the orientation of the sky and then use the diagram in the Astronomical Almanac in the Uranus (or Neptune) satellites section to determine the planet's orientation in relation to the sky orientation. Software such as Uranus viewer 2.2 can also be used in determining the location of the planet's poles. Figure 6.3 shows the orientation of Uranus and

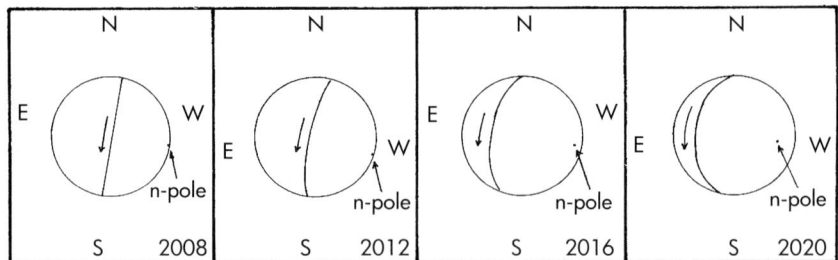

Figure 6.3. Location of Uranus's north pole (n-pole) in comparison to sky directions N = north, E = east, S = south and W = west for the years 2008, 2012, 2016 and 2020. Please note that the orientation will not change much during a year since it takes about 84 years for Uranus to make one trip around the Sun. In this figure, I have used a lower-case n to denote the planet's north pole and an upper-case N to denote the north direction in the sky. (Credit: Richard W. Schmude, Jr.)

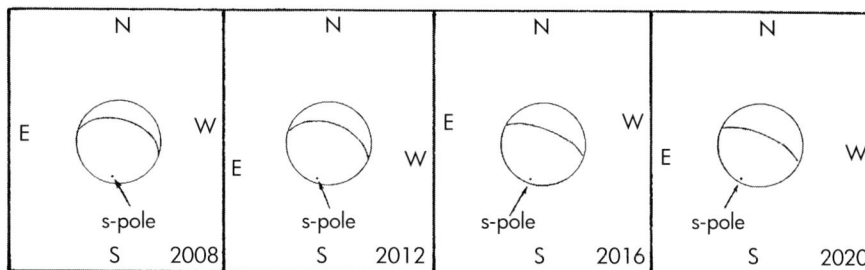

Figure 6.4. Location of Neptune's south pole (s-pole) in comparison to sky directions N = north, E = east, S = south and W = west for the years 2008, 2012, 2016, and 2020. Please note that the orientation will not change much during a year since it takes about 165 years for Neptune to make one trip around the Sun. In this figure, I have used a lower-case s to denote the planet's south pole and an upper-case S to denote the south direction in the sky. (Credit: Richard W. Schmude, Jr.)

Figure 6.4 shows the orientation of Neptune with respect to the N, S, E and W sky directions for the years 2008, 2012, 2016 and 2020.

One can also determine the location of the poles of a planet by taking an overexposed image of the planet with at least one of its brighter satellites. One can then use the satellite position(s) and the predicted position(s) in the Astronomical Almanac to determine the planet's orientation.

Visual Studies

Uranus and Neptune are difficult to draw because of their great distance and their low surface brightness. Uranus and Neptune are around 19 and 30 au from Earth and the Sun. For a comparison, Mars is seldom more than 1.6 au from Earth or the Sun. These extreme distances mean that both planets appear small even at high magnifications. In fact, Neptune at 350 × appears no larger than Jupiter through 20× binoculars. The surface brightness is the amount of light reflected by an object per unit area. Table 6.1 lists the surface brightness values of the planets and a couple of bright moons on opposition date in 2007 (Mars through Neptune) and at greatest elongation in 2007 (Mercury and Venus). The last column in Table 6.1 lists the surface brightness in terms of Neptune. Essentially 1.0 square arc-second of Venus at greatest elongation gives off 1585 times as much light as 1.0 square arc-second of Neptune at opposition. Another way of putting it is that a 0.001 second exposure of Venus will produce the same image intensity as a 1.585 second exposure of Neptune. The low surface brightness of Uranus and Neptune means that there is less light to observe. Even though Io (a moon of Jupiter) has only a third of the angular diameter of Uranus, it reflects more light. A highly magnified view of Io is easier to study than a highly magnified view of Uranus or Neptune. The lower surface brightness of the remote planets also means that color filter studies are more difficult. Filters block some light and consequently the observer has an even smaller amount of light with which to work.

If an albedo feature developed on Uranus, the likelihood of observing it would depend on several factors including seeing, transparency, telescope characteristics, observer skill, the nature/location of the albedo feature and whether the observer used one or two eyes. I will spend some time describing each of these

Table 6.1. Surface brightness values of the planets and bright moons on opposition date in 2007 (Mars through Neptune) and at greatest elongation (Mercury and Venus)

Planet	Surface Brightness (Magnitude/square arc-second)	Surface Brightness (Neptune = 1.0)
Mercury	3.5	229
Venus	1.4	1585
Mars	4.1 [a]	132 [a]
Jupiter	5.4	40
Io	5.2 [a]	48 [a]
Ganymede	5.6 [a]	33 [a]
Saturn	6.7 [b]	12 [b]
Uranus	8.4	2.5
Neptune	9.4	1.0

[a] Does not include the opposition surge.
[b] Does not include the rings.

factors. After these factors are described, I will give a few hints on how to make a drawing and then I will discuss a few successful observations of Uranus and Neptune.

Seeing

The seeing is often the biggest problem in high-magnification studies of the remote planets. All light coming from another planet must travel through Earth's atmosphere. The atmosphere contains gases that are at different temperatures and pressures. This creates air currents that are constantly moving and, since they have different temperatures and pressures, light travels through them at different speeds. The end result is that the light reaching the observer is not in exactly the same focal position. This causes the image to be blurry. Bad seeing refers to a blurry image and good seeing refers to a steady and sharp image. Under excellent seeing conditions, one can increase his or her magnification to approximately 50× per inch of telescope aperture. As the altitude of the target rises, the seeing gets better because the light travels through a thinner part of the atmosphere. Problems associated with poor seeing come from the local surroundings and air currents.

There are some things that one can do to improve seeing. First, there should be no warm or hot objects near the optical path. Heat will cause air turbulence and poor seeing. One should stay away from buildings, pavement and paved roads when observing. These items absorb heat during the day and release it slowly at night causing extra air turbulence and poorer seeing. Grass and wood do not absorb as much heat and will yield better seeing at night. If one uses a concrete telescope pad it should be painted white. White paint reflects sunlight and, as a result, the pad will not absorb as much heat.

There are three ways of determining seeing. In the first method, seeing is rated on a scale of 0 (poor) to 10 (perfect). This scale is used by members of the ALPO. Very recently, word descriptions were given to each number; more information about this

Table 6.2. Seeing scale developed by Eugene Antoniadi

Seeing	Description
I	Excellent conditions, image is steady even at the highest magnification
II	Presence of occasional atmospheric turbulence, but with moments of calm lasting several seconds
III	Frequent atmospheric turbulence permitting medium powers to be used
IV	Poor conditions with nearly constant episodes of atmospheric turbulence and detail can only be seen occasionally
V	Very poor conditions, with a blurred image even at low power

is in *Saturn and how to Observe it* ©2005 by Julius Benton. A second method – developed by Eugene Antoniadi – is summarized in Table 6.2. This scale is often used in Europe. A third method is to estimate the size of a star. For example, if a star appears to have an angular diameter of 2.0 arc-seconds, we say that the seeing is 2.0 arc-seconds. (Note: one can also image a star and measure its size.)

One method that I have developed to estimate the seeing is to observe Saturn and estimate how far the Cassini Division is visible. If the gap is completely visible, this would indicate excellent seeing. (The Cassini Division is the dark gap between Saturn's bright outer A ring and its bright inner B ring.) The advantages of this scale are (1) it is based on observing a planet instead of a star, (2) it is quick to use and (3) one can compute the exact width of the Cassini Division.

Transparency

Transparency is a measure of the amount of light that is getting through our atmosphere. While a thin haze may help one get a better view of a bright planet like Mars, it is a problem for the remote planets because of their low surface brightnesses. Therefore, recording the transparency is important when reporting visual observations of Uranus and Neptune. There are two ways of estimating the transparency.

One method is to determine the magnitude of the faintest star which is at the same altitude as the target. This star should be at the limit of visibility in direct vision. The problem with this method, however, is that one may not find a star meeting such criteria.

A better method of estimating the limiting magnitude is to determine the faintest star that is visible and the brightest star that is not visible. This may seem confusing but bear with me. The first two columns in Table 6.3 list stars for a polygon outlined by stars in the constellation Pisces. See Figure 6.5A. The third and fourth columns in this table list stars for a polygon outlined by stars in the constellations Pisces and Aries. See Figure 6.5B. One simply counts the number of stars that are visible inside of the polygon using direct vision along with those that define the polygon and then reads off the magnitude value. This value is the faintest magnitude observable. The magnitude just below is the brightest star that was not seen. The limiting magnitude is probably between these two values. For example, let's say that one uses the first two columns in Table 6.3 and counts

Table 6.3. Tables for determining the limiting magnitude for two areas in the sky near Uranus and Neptune

Lambda(λ), Iota(ι), Theta(θ), Beta (β) and Kappa(κ) Piscium	Magnitude	Kappa(κ), Alpha(α), Lambda(λ), and Beta(β) Arietis, Eta(η) and Omicron (o) Piscium	Magnitude
Number of stars seen in polygon		Number of stars seen in polygon	
1	3.9	1	2.2
2	4.2	2	2.7
5	4.5	3	3.8
6	5.0	4	3.9
7	5.3	5	4.5
8	5.8	6	4.8
9	6.7	7	5.1
11	6.9	8	5.3
		9	5.8
		11	6.2
		12	6.6

seven stars, which includes the stars making up the polygon. Note that the first column lists the number of stars seen and the second column lists the limiting magnitude. The faintest star seen is magnitude 5.3 and the brightest star not seen is magnitude 5.8 and so the limiting magnitude is in between these values at 5.55 (or rounded up to 5.6). I used Equation 4.1 along with b = 0.21 to compute the magnitude values in Table 6.3.

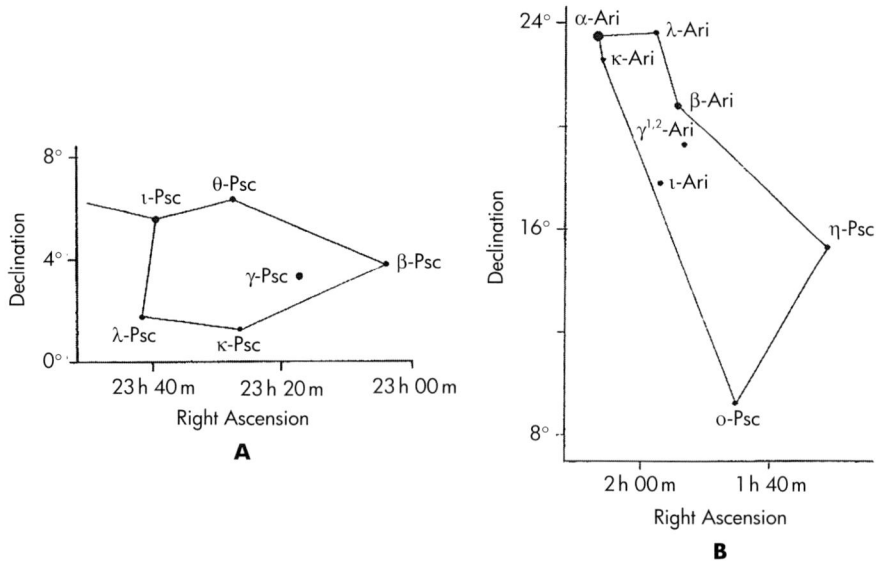

Figure 6.5. Figures described in table 6.3 that can be used in estimating the limiting magnitude. (Credit: Richard W. Schmude, Jr.)

Telescope Characteristics

The telescope and its characteristics will affect the view of the remote planets. Three important characteristics which I would like to discuss are the telescope diameter, collimation and Barlow lenses.

The amount of light coming through the telescope is directly proportional to the area of the main mirror or objective lens. It is important for the observer to have as much light as possible enter his/her system when studying the remote planets. The resolution also improves with larger diameter telescopes. (More information about the resolution and light gathering power of telescopes is contained in Chapter 4.) Since there are so many things that can affect whether an albedo feature is noticed, it is difficult to set a lower limit to the size of telescope needed to record features. My feeling is that, for most people, a diameter of at least six inches for refractors and 8 inches for other telescopes is needed to see the most obvious irregularities on Uranus. Larger telescopes are needed for Neptune.

Collimation is very important for making drawings and images of the remote planets. A well collimated telescope will yield a clear and sharp focus whereas a poorly collimated one will not. Collimation deals with the alignment of the optical surfaces within a telescope. As a general rule, the higher the f-number of a Newtonian, the easier it is to collimate. If one uses a portable telescope that is not a refractor, they should check the collimation before each observing run. There are some who would disagree with this and say that a collimation check before each observing session is not needed. Keep in mind though that even a slight collimation problem will be obvious at the high magnifications needed for the remote planets. If, however, one is more interested in examining deep sky objects at low magnifications, a slight collimation problem will not degrade seriously the view. For permanently mounted telescopes, the collimation should be checked periodically. One should follow the telescope manufacturer's instructions on proper collimation.

Modern Barlow lenses will help one see detail on the remote planets. This is because they magnify the image and preserve the eye-relief of the eyepiece. Generally the eye-relief decreases with increasing magnification. This lens is of great value when using a telescope with a low f-number. In spite of the additional optical surfaces in a Barlow lens, the loss of light is only about 1% if the lens has the proper coatings.

Other factors that should be checked include the state of the optical coatings, the mirror and stray light. An open tube telescope can suffer from stray light which can degrade the view. Telescopes should also have a dew-shield, which will block out stray light and also prevent the formation of dew.

Each Newtonian telescope has a secondary mirror, which is attached to it by metal rods, called spider vanes. The spider vanes lie in the optical path and can degrade the image. One way around this problem is to create an opaque mask which fits in front of the telescope. The mask should contain a hole which is small enough to fit between the spider vanes. I tried this once with Mars and had an excellent view. The obvious problem with this approach is that it reduces the amount of light passing through the telescope. One should also use as small of a secondary mirror as possible to improve telescope performance. It is easier to use a small secondary (less than 20% of the main mirror diameter) on instruments with a large f-number than with small f-numbers.

Observer Skill

The skill of the observer plays a crucial role in whether an albedo feature will be observed. As it turns out, under low light levels, the eye is most sensitive to blue-green light. Therefore, the color of a feature may determine whether it is detected or not. If an observer has astigmatism, he/she should compensate for that by wearing corrective lenses. The observer's experience is the most critical factor in the detection of low-contrast albedo features on Uranus and Neptune.

The Nature/Location of the Albedo Feature

The nature/location of the albedo feature is critical to determining its visibility. The nature of a feature includes its size and how it interacts with light. A large feature will be easier to notice than a small one. A feature can only be seen by our eyes if it affects the amount of visible light reflected. Several leading astronomy magazines have published images of Uranus and Neptune showing clouds and belts. Most of these images, however, were made in near-infrared light. Many features reflect near-infrared light but not visible light; hence, they can only be imaged in wavelengths which are invisible to the human eye.

A few amateurs have begun taking infrared images and I feel that these images will yield fruitful results. Many CCD cameras are sensitive to near-infrared light; however, Uranus and Neptune are dim in this light. If one wants to image these planets in near-infrared light, he or she should use a large telescope with a good mount, make long exposures and use filters which block out visible light.

If a notable feature is seen, one should get a friend to look for it. He or she should wait about 45 minutes and look again to see if it has changed position.

Observing With One or Two Eyes

There are several advantages of using both eyes when observing the remote planets. One is that low contrast features are easier to see with two eyes than with one eye. Try looking at a distant and faint object with one eye, and then use two eyes. The difference should be obvious. A second advantage is that when two eyes are used there is no interference from an unused eye. One can also detect fainter objects and improve their resolution by using both eyes. Finally, when both eyes see an object, any defect(s) in an eye(s) is/are reduced or eliminated by the brain.

There are two ways of using both eyes for viewing, namely, a binocular eyepiece mount and a binocular telescope. There are several vendors who sell binocular eyepiece mounts. If possible, insist on a model that has coated optics. There should also be a way to adjust the mount to accommodate the distance between one's eyes; otherwise this may cause eyestrain. Finally, two identical eyepieces are needed.

A binocular telescope is an ideal instrument for observing the remote planets provided that one can attain a magnification of at least 300X. At least one company (JMI) sells a 0.4 m (16 inch) binocular telescope, which is capable of high magnifications. Some have also built their own binocular telescopes. See, for example, Sky & Telescope, Feb. 1993 p. 89. One advantage of using a binocular telescope is that, since two telescopes are used, the brain averages the seeing. Like Newtonians, one should check the collimation of their high magnification binoculars before an observing session.

Drawing Hints

In this section, I will give some tips for drawing Uranus and Neptune. Before making a drawing, one should allow his or her telescope to reach ambient temperature. Generally, this will take 15 to 20 minutes. One should have a hard surface on which to write such as a clip-board, together with a pencil, an eraser, a clip-on red light, a clock and an ALPO observing form. A copy of an ALPO observing form is in the Appendix. All blanks on the form should be filled in. When observing Uranus or Neptune, one should select a night when the seeing is good to excellent (I or II on the Antoniadi scale).

After finding the object, one should use a magnification that is high enough to show an adequate sized disc, but low enough to preserve good definition. One should refocus the telescope every minute or so. This has helped me see more planetary detail. When making a drawing, be patient with the seeing since every once in a while the atmosphere will settle down and faint shadings may appear. Watch to see if the shadings remain consistent or if they jump around the disc. Consistent shadings are probably real features. On many occasions when viewing Mars and Jupiter, I saw the most detail in the first ten minutes of the observing session.

The first feature to look for on Uranus and Neptune is limb darkening, which is the darkening near the edges. After drawing the limb darkening look for any differences in the limb darkening along the circumference of the disc, and, if an irregularity is seen, note its location. Afterwards look for any polar flattening and note it. Finally, scrutinize the disc and look for additional irregularities or bands. Because of their smaller angular sizes, Uranus and Neptune require less time to draw than Jupiter. After about 15 minutes at the eyepiece, determine the orientation of the drawing by noting the N, S, E and W directions of the sky, but be sure to note that sky directions are used. There have been many instances when people have submitted drawings with north and east directions but they did not indicate whether these directions were of the sky or of the planet.

Sample Observations

In this section, I would like to present some sample visible-light drawings of Uranus. A few of these are shown in Figure 6.6. Stephen J. O'Meara made the first drawing in Figure 6.6 on September 15, 1981, with a 0.23 m (9 inch) refractor. He observed a bright cloud on Uranus, which moved as that planet rotated. He was able to determine a rotation period of ~16.4 hours for Uranus by watching this bright cloud. Four years later, Voyager 2 data yielded an almost identical rotation period for that planet. The writer made the second drawing in the top row in Figure 6.6 on July 15, 1988. I used the 0.36 m (14 inch) Schmidt-Cassegrain telescope at Texas A&M University Observatory to observe Uranus. The

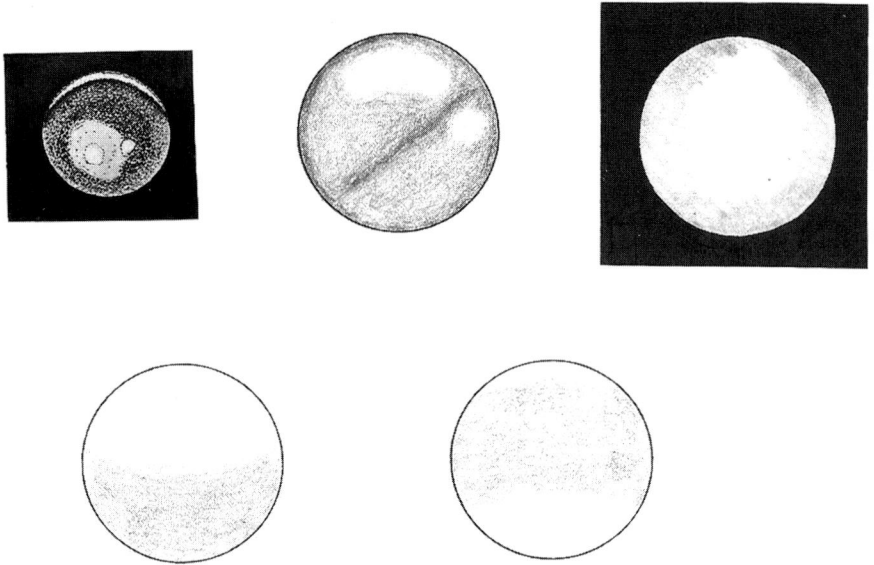

Figure 6.6. Drawings of Uranus. In all cases, the planet's north pole either is at the top or is above the center of the disc. Top left: Stephen J. O'Meara, Sept. 15, 1981 (0:00 UT), 0.23 m (9 inch) refractor at Harvard College Observatory. Top Center: Richard W. Schmude, Jr. July 15, 1988 (3:48 UT), 0.36 m Schmidt-Cassegrain at Texas A&M University Observatory, 530X, seeing was excellent. Top Right: Detlev Niechoy, Sept. 23, 2006 (21:47 UT), 0.20 m Schmidt-Cassegrain, 340X, seeing = II to III on the Antoniadi scale. Bottom left: Phil Plante, Oct. 15, 2006 (2:49 UT), 0.64 m Newtonian, 320X, seeing = III on the Antoniadi scale. Bottom right: Richard Jakiel, Nov. 10, 2006 (2:30 UT), 0.30 m (12 inch) Schmidt-Cassegrain, 510X, seeing was excellent. (Credit: Stephen J. O'Meara, Richard W. Schmude, Jr., Detlev Niechoy, Phil Plante and Richard Jakiel)

magnification was 530× and the seeing was excellent. A bright area towards the west direction (sky direction) was obvious; Uranus's pole was near the center of the disc. Two others independently confirmed this feature while a third person reported a color change in this area. Detlev Niechoy, a German amateur, made the third drawing of Uranus. The most important feature here is the weaker limb darkening near Uranus's northern limb. Phil Plante made the bottom left drawing in Figure 6.6 on October 15, 2006. The most important feature here is that Uranus's northern hemisphere is brighter than the southern hemisphere. Richard Jakiel made the bottom-right drawing in Figure 6.6 on Nov. 10, 2006 at 2:32 UT. He used a 0.30 m (12 inch) cassegrain telescope at a magnification of 510× under excellent seeing conditions. He saw a large bright south polar region and a smaller bright north polar region.

Figure 6.7 shows three drawings of Neptune. The writer used the 30 inch reflector telescope on top of Freemont Peak in California to make the left drawing in Figure 6.7. I made it on July 19, 1992 (7:10 UT) using magnifications of 230× and 370×. The seeing and transparency were good. The only irregularity was strong limb darkening; the disc had a slight blue color and its edge was sharp. Brian Cudnik made the center drawing on Oct. 17, 2004 with a 0.36 m (14 inch) Schmidt-Cassegrain telescope. The seeing was excellent. He described the bright area just south of the center as "seemed present but with vague definition". He described the

Figure 6.7. Drawings of Neptune. In all cases, the planet's north pole is at the top. Left: Richard W. Schmude, Jr., July 19, 1992 (7:05 UT), 0.76 m Newtonian, 370X, seeing = 7 on the ALPO scale. Center: Brian Cudnik, Oct. 17, 2004 (0:48 UT), 0.36 m (14 inch) Schmidt-Cassegrain, seeing = 7 to 9 on the ALPO scale. Right: Brian Cudnik, Oct. 9, 2005 (5:04 UT), 0.36 m (14 inch) Schmidt-Cassegrain, 490X, seeing = 5 to 7 on the ALPO scale. (Credit: Brian Cudnik and Richard W. Schmude, Jr.)

disc as having a blue-green hue. Brian also made the third drawing in Figure 6.7. He made it on Oct. 9, 2005 (5:00 to 5:08 UT) with a 0.36 m (14 inch) Schmidt-Cassegrain telescope at 490×. The seeing conditions were good and the transparency was excellent. He used W21 (orange) and W25 (red) filters to make this drawing. He noted an elongated bright spot near the center of Neptune along with some limb darkening.

In addition to the drawings in Figures 6.6 and 6.7, others have also succeeded in making observations of Uranus and Neptune. I will describe a few of these.

Eugene Cross, Jr. and Randy Shartle observed dark belts on Uranus during the mid to late 1960 s. During this time, Uranus's equator was facing the Earth. E. M. Antoniadi also observed dark belts on Uranus with a 0.84 m (33 inch) refractor when that planet's equator faced the Earth in the 1920s.

William Sheehan and Stephen J. O'Meara used the 1.0 m f/16 Cassegrain telescope at Pic Du Midi Observatory in 1992 to observe both Uranus and Neptune. They were not able to see distinct albedo features on Uranus but were able to see several features on Neptune. The most distinct feature was a dark belt near the southern limb. In addition, these two noticed a similar dark belt near the northern limb along with a few light and dark shaded areas between the two polar regions. They also saw a small bright cloud near Neptune's southern limb on one of the nights.

Norman Boisclair made two important negative observations on Oct. 4, 2005, under excellent seeing conditions. He used a 0.5 m (20 inch) Newtonian telescope at magnifications of 1000× (Uranus) and 840× (Neptune). He noticed sharp limbs and limb darkening on both planets but no other irregularities. He reported that Neptune's limb darkening was more obvious than Uranus's limb darkening.

It is important to make drawings under good seeing with a high quality telescope. Negative observations are also valuable and should be sent to the ALPO and BAA (British Astronomical Association) remote planets coordinators.

Ellipticity Studies

All rotating bodies experience centripetal force. This force causes a body to bulge near its equator and have a non-circular shape. The size of the visible bulge depends on several factors including the composition of the rotating body, its rotational rate and the transparency of its atmosphere. The ellipticity defines how much a body's shape deviates from a sphere.

The ellipticity (ε) is computed from:

$$\varepsilon = (E_D - P_D)/E_D \qquad (6.1)$$

where E_D is the equatorial diameter and P_D is the polar diameter. Astronomers used Voyager 2 data to measure ellipticity values of 0.0229 for Uranus and 0.017 for Neptune. Since the tilt of Uranus and Neptune changes, the apparent value of as seen from the Earth is usually less than these values. This is because the P_D value appears larger than what it is in actuality.

Tables 6.4 and 6.5 show the relationship between the sub-Earth latitude and the apparent polar diameter of Uranus and Neptune respectively. The brightness increase (ΔI), as a result of the sub-Earth latitude being far from Uranus's equator, is computed from:

$$\Delta I = 2.5 \times \log(D_P{}'/D_P) \qquad (6.2)$$

where $D_P{}'$ is the apparent polar diameter as seen from Earth and D_P is the true polar diameter which is seen only when the sub-Earth latitude $= 0°$. The brightness increases are due to the larger geometrical size of the disc caused by the observer seeing $D_P{}'$ instead of D_P. This is shown in Figures 5.7 and 5.8.

R. B. Minton reported in the early 1970 s that Jupiter's equatorial diameter in methane band light (wavelength = 890 nm) was 1.3% smaller than expected. He also explained that this is what causes Jupiter to have a nearly circular shape in methane band images and thus a different ellipticity than observed at visible wavelengths. The writer has also found evidence that Jupiter's ellipticity changed

Table 6.4. The apparent polar diameter, ellipticity and magnitude increase for different sub-Earth latitudes for Uranus

Sub-Earth latitude(°N or °S)	Apparent polar diameter(km)	Apparent ellipticity	Brightness increase (magnitudes)
90	51,118	0.0000	0.025
82.23	51,095	0.0005	0.025
80	51,081	0.0007	0.024
70	50,977	0.0028	0.022
60	50,818	0.0059	0.019
50	50,625	0.0096	0.015
40	50,422	0.0136	0.010
30	50,233	0.0173	0.006
20	50,081	0.0203	0.003
10	49,982	0.0222	0.001
0	49,948	0.0229	0.000

Table 6.5. The apparent polar diameter, ellipticity, and magnitude drop for different sub-Earth latitudes for Neptune

Sub-Earth latitude(°N or °S)	Apparent polar diameter(km)	Apparent ellipticity	Brightness increase (magnitudes)
28.33	48,875	0.0132	0.004
20	48,779	0.0151	0.002
10	48,706	0.0166	0.001
0	48,682	0.0171	0.000

in ultraviolet light in 2001–2002. Similar changes in the ellipticity at infrared and ultraviolet wavelengths of Uranus and Neptune should be searched once images, at suitable resolution, become available.

Polarization Studies

Polarization data can yield information about a planet's atmosphere, chemical composition, and the mean size and shape of haze particles. Polarization data can also yield information about the soil particles for bodies with solid surfaces.

Two professional astronomers report polarization values for Uranus's entire disc of 0.5 and 0.3 polarization units for blue and red light respectively during 1975–1976. Polarization units are discussed later in this chapter. From these results, they concluded that haze or a thin, high-altitude cloud was present. More recently, a second group reports that the amount of polarized light near the limbs of Uranus and Neptune can reach ∼1%. This group also reports that the amount of polarized light increases with decreasing wavelength of light for both planets. They also point out that the distribution of polarized light on Uranus and Neptune is different than on Jupiter. This is consistent with the haze layer being different on Jupiter than on Uranus and Neptune. Many questions remain to be answered such as; Does the polarization value change as a cloud passes? Does the polarization value change at different seasons? What is Pluto's polarization value and does it change as it rotates?

One may wonder how to collect polarization data. Before we answer this question, we will discuss a few characteristics of light and polarizing filters. This will be followed by a discussion of how polarization measurements are made. Finally, an example is worked out.

Electromagnetic radiation contains two waves that are at right angles to one another, namely, an electric wave and a magnetic wave. See Figure 6.8. If we look at an electromagnetic wave straight on it will look like a plus sign with the electric wave moving in one plane and the magnetic wave moving in a plane which is perpendicular to it. In Figure 6.9, the electric wave is moving in a vertical plane and the magnetic wave is moving in a horizontal plane. Light that is strongly-polarized is oriented in the same direction as is shown in Figure 6.10, whereas non-polarized light has random directions similar to what is shown in Figure 6.11. Light which is polarized-partially constitutes a combination of polarized and non-polarized light similar to what is shown in Figure 6.12.

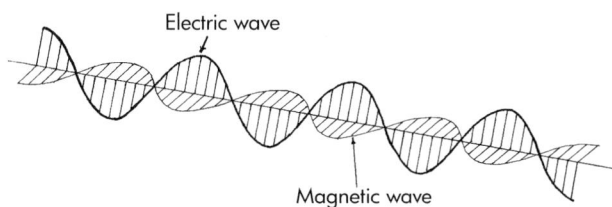

Figure 6.8. A diagram showing the two parts of an electromagnetic wave. The electric wave moves in a plane that is perpendicular to the plane in which the magnetic wave moves. (Credit: Richard W. Schmude, Jr.)

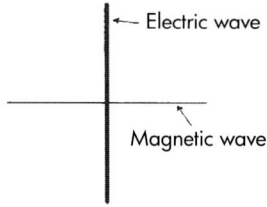

Figure 6.9. A highly magnified view showing how light looks as it moves directly at the viewer. The bold line is the plane that contains the electric wave and the thin line is the plane that contains the magnetic wave. (Credit: Richard W. Schmude, Jr.)

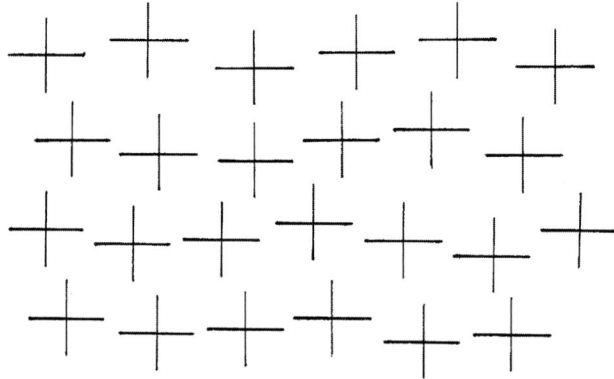

Figure 6.10. A highly magnified view showing how polarized light looks as it approaches the viewer. In all cases here, the electric waves move in planes that are in a horizontal direction and the magnetic waves move in planes that are in a vertical direction. (Credit: Richard W. Schmude, Jr.)

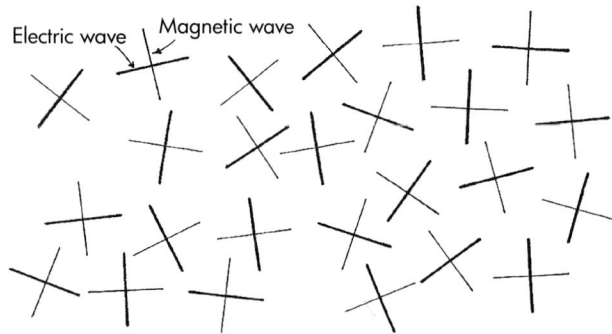

Figure 6.11. A highly magnified view showing how non-polarized light looks as it approaches the viewer. The orientations of the planes containing the electric and magnetic waves are in random directions. (Credit: Richard W. Schmude, Jr.)

Polarization data is collected with a polarizing filter, which contains millions of long molecules aligned in the same direction similar to what is shown in Figure 6.13. The polarizer prevents light which has an electric field parallel to the molecules to pass through. If one rotates a polarizer and examines polarized light, he/she will notice a light intensity change. This is due to the fact that the polarizer blocks light with an electric wave which is in the same plane as the molecules. If, on the other hand, the light is non-polarized, the light intensity will

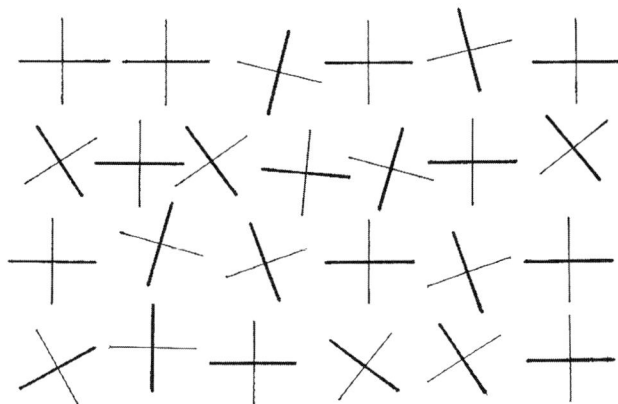

Figure 6.12. A highly magnified view showing how partially-polarized light looks as it approaches the viewer. A larger number of photons than expected from random orientation have electric waves moving in a horizontal direction. (Credit: Richard W. Schmude, Jr.)

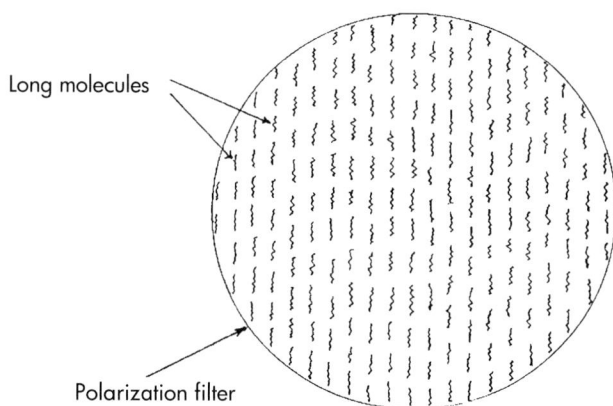

Long molecules

Polarization filter

Figure 6.13. A highly magnified view of part of a polarizing filter showing the orientation of its long molecules. Each small squiggle represents a long molecule. These molecules only allow light with certain orientations to pass through. (Credit: Richard W. Schmude, Jr.)

not change as the polarizer is rotated. Table 6.6 lists examples of non-polarized, partially-polarized and strongly-polarized light.

One can obtain polarization data with a telescope, polarizer and a light detector such as a photoelectric photometer. One uses equation 6.3 to compute the degree of polarization (P)

$$P = 1000 \times (I_P - I_R)/(I_P + I_R) \qquad (6.3)$$

where I_P is the intensity perpendicular to the plane of vision and I_R is the measured intensity in the plane of vision. The plane of vision is defined by three points, namely, the observer, the target and the source of illumination, which is the Sun for the remote planets. The plane of vision can be estimated for Uranus and Neptune by looking for a bright star or a planet that lies close to the ecliptic, which is the path that the Sun appears to move in across the sky. One then orients the polarizer accordingly. One must have a sensitive detector and a large telescope

Table 6.6. Examples of non-polarized, partially-polarized, and strongly-polarized light

Non-polarized light	1. Light from an incandescent bulb
	2. Light from a fire
Partially-polarized light	1. Light scattered by the blue sky
	2. Light reflected by a first quarter moon
Strongly-polarized light	1. Light reflected by many car windows
	2. Light coming through a polarizer

to make polarization measurements of the remote planets; furthermore, the polarizer must be made out of a rigid material like glass. Large plastic polarizers have a tendency to bend and if this occurs, error will be introduced into the measurements.

One interesting measurement that can be made is to measure the amount of polarized light that the polar regions of Uranus, Neptune, and Pluto reflect, and to compare this to what the equatorial regions reflect. It would be interesting also to see how a great dark spot on Neptune affects the amount of polarized light reflected by that planet. One can also measure the amount of polarized light that Pluto reflects as it rotates. There is a chance that some areas reflect more polarized light than other areas, and this can give us clues about the nature of Pluto's surface. Polarization data may also yield information about Pluto's atmosphere. One can also measure how the amount of polarized light changes with the solar phase angle.

Example of a Polarization Measurement

I have worked out an example of a recent polarization measurement of Mars. The method is exactly the same for Uranus. The polarization data was collected on March 4, 2006, near Barnesville, Georgia, and the data and analysis are summarized in Table 6.7. In this example, the polarizer was aligned along a line defined by Mars and Saturn. The plane of vision was defined by Saturn, Mars and the observer.

I started the polarization measurement by moving my telescope so that the photometer FOV was just right of Mars and oriented the polarizer so that its plane was in the plane of vision. Then I took three polarization readings (782, 784 and 784) in the first row of data under the Sky Readings (right) column in Table 6.7 (parallel). After this, I moved my telescope so that Mars was at the center of the photometer FOV and took the six readings, which are in the first row of data under the Mars + Sky Readings column (1025, 1025, 1028, 1023, 1020 and 1023) in Table 6.7. Finally, I moved the telescope until the photometer FOV was just left of Mars and took three more sky readings (788, 789 and 790) under the Sky Readings (left) column in Table 6.7. I changed the position of the photometer FOV in relation to Mars three times as I took the data in the first row of Table 6.7. These position changes are shown in Figure 6.14. I recorded sky measurements because scattered light from the sky are often polarized.

After I took the data in the first row in Table 6.7, I rotated the polarizer 90° and repeated this routine. The readings for this orientation of the polarizer are shown in the second row of data in Table 6.7 (perpendicular).

Table 6.7. Polarization data of Mars collected by the writer on March 4, 2006, between 2:01 and 2:24 UT

Orientation	Sky Readings(right)	Mars + Sky Readings	Sky Readings(left)	I_P or I_R
Parallel	782, 784, 784	1025, 1025, 1028, 1023, 1020, 1023	788, 789, 790	237.833
Perpendicular	756, 759, 758	1002, 999, 1015, 1001, 1000, 1005	755, 755, 750	248.167
Parallel	780, 783, 773	1006, 1004, 1003, 1003, 1002, 1001	762, 762, 761	233.000
Perpendicular	730, 737, 732	976, 976, 975, 977, 974, 974	729, 725, 726	245.500
Parallel	758, 759, 759	996, 993, 997, 995, 997, 995	760, 760, 759	236.333
Perpendicular	727, 726, 727	973, 968, 968, 970, 969, 968	722, 724, 724	244.333
Parallel	757, 756, 756	992, 991, 989, 988, 989, 986	750, 752, 750	235.667
Perpendicular	721, 715, 719	963, 963, 963, 963, 964, 961	717, 718, 716	245.167
Parallel	748, 747, 749	982, 979, 978, 981, 980, 979	745, 749, 746	232.500

I continued making measurements until there were a total of five sets of parallel orientations and four sets of perpendicular orientations. I recorded a new row of data for each orientation of the polarizer.

I computed I_R by first computing the average Mars reading (r_M) and the average sky reading (r_S) as:

$$r_M = 1024.00$$

$$r_S = 786.167$$

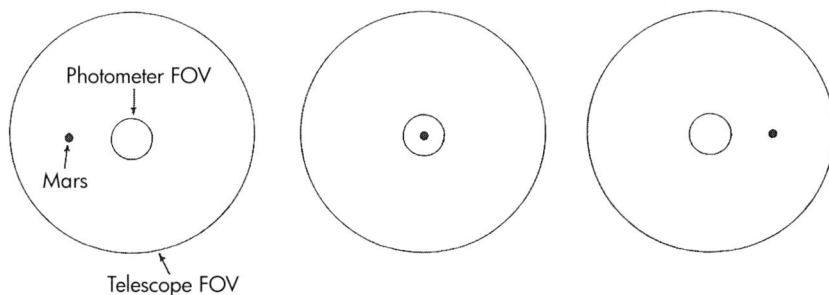

Figure 6.14. One must make three different measurements with each orientation of the polarizing filter. In all cases, the small central open circle is the photometer FOV. First one measures the amount of scattered light that is polarized by the sky to the right of the target (left frame); then measures the same quantity for the target (center frame) and then measures the same quantity for the sky to the left of the target (right frame). This sequence is repeated for different orientations of the polarizer. (Credit: Richard W. Schmude, Jr.)

The first I_R value is

$$I_R = r_M - r_S = 237.833. \text{ One then computes the first } I_P \text{ value}$$

$$I_P = r_M - r_S = 1003.667 - 755.500 = 248.167.$$

I computed the remaining I_P and I_R values in the same way. The results are listed in the last column in Table 6.7.

I used equation 6.3 to compute the degree of polarization. Since an I_R value was measured immediately before and after I_P, I used an average value for I_R for each I_P measurement.

$$I_R = (237.833 + 233.000)/2 = 235.417$$

Then I computed the first polarization value as:

$$P = 1000 \times (248.167 - 235.417)/(248.167 + 235.417)$$

$$= 1000 \times (12.75/483.584) = +26.37.$$

Please note the positive sign. As it turns out, the polarization value can be positive or negative.

I computed the second polarization value as:

$$I_P = 245.50$$

$$I_R = (233.00 + 236.333)/2 = 234.667$$

$$P = 1000 \times (245.50 - 234.667)/(245.50 + 234.667)$$

$$= 1000 \times (10.833/480.167) = +22.56.$$

The third and fourth polarization values are $+17.35$ and $+23.13$. Then I computed an average polarization value of $+22.4$.

Methane Band Imaging

Methane band images are made with an electronic camera and a methane band filter. A methane band filter is one that allows in light which is absorbed by methane. One popular methane band filter has a peak transmission near 890 nm. Since many methane filters have narrow FWHT values, a large telescope is needed to make useful images with them. Methane band images can be used in detecting high altitude clouds on Uranus and Neptune. Figure 6.15 illustrates how this is accomplished. Essentially light with a wavelength of 890 nm (or some other wavelength which methane absorbs) is absorbed by the atmospheric methane. The only methane light which is reflected back is light that bounces off a high altitude cloud. A high altitude cloud will appear bright in a methane band image. The small amount of methane in Earths atmosphere has almost no effect on these images.

Methane band photometry can yield information on outbursts of cloud development. This type of photometry is especially sensitive to high altitude clouds and hazes. There are two ways of carrying out methane band photometry. The first method is to extract photometric data from digital images. The advantage here is that data can be collected for either part of the planet or its entire disc. The second method is to place a methane band filter in front of a photoelectric photometer.

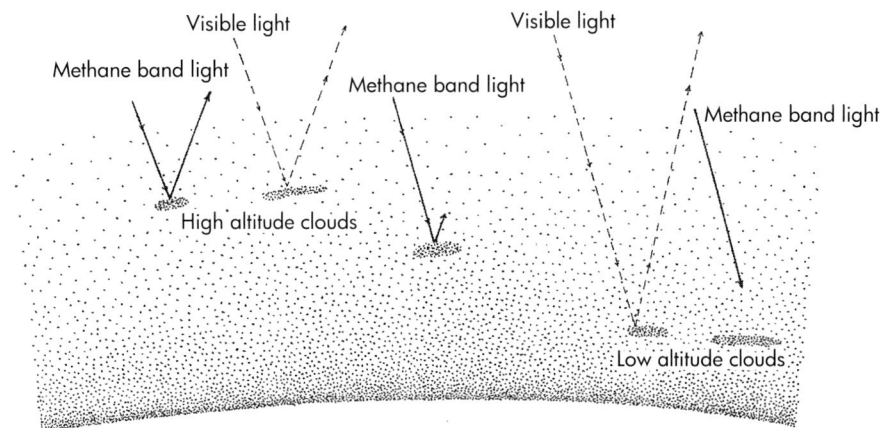

Figure 6.15. Methane band light is absorbed strongly by the methane in Uranus's (and Neptune's) atmosphere, whereas visible light is not absorbed. Therefore, bright objects in a methane band image will have a high altitude because the only methane band light that is reflected back to Earth is that which is reflected by high altitude clouds. Methane band light is unable to penetrate the lower parts of Uranus's atmosphere. (Credit: Richard W. Schmude, Jr.)

Since Uranus and Neptune are dim at 890 nm, one will need a sensitive detector and a large telescope to carry out methane band photometry.

Professional astronomers carried out methane band photometry of Neptune in the late 1980s. These results show that Neptune underwent a substantial brightness increase in 1987 but it subsided by 1990. The brightness change was caused by the appearance and dissipation of high altitude clouds on that planet.

Imaging

Before imaging, one must check the collimation of his or her telescope. Willem Kivits, a Dutch amateur astronomer, has shown that if the telescope collimation is not perfect, a bright limb spot can show up in images of Uranus after processing. Therefore, telescope collimation must be near perfect before one can image Uranus. If one images an albedo feature, they should rotate his or her camera, take a second image and repeat the same processing routine and check to see that the feature did not move with the camera.

One must also be on guard for Earth's atmosphere acting as a prism. Essentially when a planet's altitude drops below about 30°, one edge of it will appear bluish and the opposite edge will appear reddish. I have seen this several times when Jupiter was less than 30° above the horizon. This effect may show up when the planet is at elevations above 30° in electronic images because CCD cameras are more sensitive than the human eye. Look for this effect by making a color image, and look for a reddish tinge on one side and a bluish tinge on the opposite side. Since most CCD cameras are more sensitive to red than blue light, the red portion may show up as a bright area in images.

When making images of Uranus and Neptune, one must give information about the orientation of the image. Sky directions are all that are needed.

Sample Images

In the next section, we will summarize two successful ways of making images of Uranus. This list is in no way complete; however, it should give the reader a starting point for making good images of Uranus and Neptune.

Don Parker recorded the Uranus image in Figure 6.16. His equipment and technique are summarized in Table 6.8. The image was made on September 8, 2006, in visible light. The most obvious feature is the brightening on the planet's southern limb. There is also a bit of limb brightening in the image which is not equal on the disc's east and west limbs. This image shows an ellipticity of 0.02.

Christophe Pellier recorded the Uranus image in Figure 6.17. His equipment and technique are summarized in Table 6.9. He used red and infrared light to make this image on June 28, 2004, at 2:54 UT. The most obvious feature here is the bright

Figure 6.16. An image of Uranus recorded by Don Parker on Sept. 8, 2006, with his 0.41 m reflecting telescope in his back yard near Miami, Florida. The details of how this image was made are summarized in Table 6.8. Sky directions are shown and Uranus' north limb is near the top. (Credit: Don Parker, ALPO Remote Planets Section.)

Table 6.8. Summary of Don Parker's equipment and technique in making the Uranus image shown in Figure 6.16

Camera	Skynyx 2-0
Telescope	0.40 m (16 inch) Newtonian at f/22 (Barlow lens was used)
Filter	LRGB (Luminance, red, green and blue all with infrared blocking)
Exposure time	94 milliseconds
Exposures taken and used	1272 exposures were taken per filter; 890, 950, 838 and 975 were used for the L, R, G and B filters respectively
Pixel size	7.4 μm (0.17 arc-seconds in the image)
Focuser	Starlight Instruments feather touch focuser with robofocus motor
Software	Lucam recorder software Registax 3, Maxim DL and Photoshop
Dark and flat field corrections	None made; Don felt that the dark current correction made little difference to the image quality
Processing	He used Registax 3 to create an L, R, G and B image from several hundred raw frames and then used Maxim DL to combine, align and enlarge the images. He then carried out a Lucy-Richardson deconvolution to sharpen the image, and finally he used Photoshop to do a color balance and enhance the contrast.

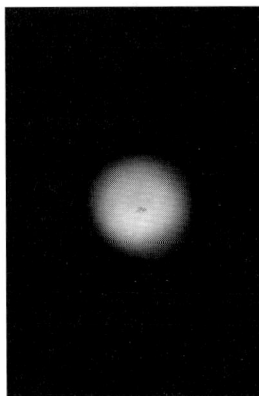

Figure 6.17. An image of Uranus recorded by Christophe Pellier with a 0.36 m Schmidt-Cassegrain telescope. The details of this image are summarized in Table 6.9. Uranus' north limb is near the top. (Credit: Christophe Pellier, ALPO Remote Planets Section.)

Table 6.9. An overview of the equipment and technique used by Christophe Pellier in making the image in Figure 6.17

Camera	ATK-1HS
Telescope	0.36 m (14 inch) Schmidt-Cassegrain
Filter	Red and infrared filters
Exposure time	0.25 seconds/frame for RG image, 0.50 seconds/frame for IR700 image
Exposures used	Several hundred were used in constructing the image
Pixel size	0.54 µm
Focuser	Original focuser on the Celestron 14 telescope
Software	Acquisition software: Qcfocus [a]; images stacked with Registax version 1
Processing	A wavelet processing routine was used to enhance details

[a] Written by Patrick Chevaley

shading in Uranus's south polar region. The image has a small amount of limb darkening and it shows an ellipticity of 0.02.

When making images, it is important to keep in mind that professional astronomers are interested in any changes that occur on Uranus or Neptune; therefore, if one images an albedo feature he/she should take a second image about an hour later and check to see if there is any change in its position. It is important that we have information about the orientation of the planet whenever an albedo feature is imaged. Professional astronomers prefer that images have dark-frame and flat-field corrections but no other processing. They also prefer that images be stored as FITS files.

Occultation Measurements

There are three classes of occultations that can yield data on Uranus, Neptune, and Pluto. The first class involves a planet, its rings or its moons occulting a more distant star. The second class occurs when Earth's moon blocks out a more distant planet. A third class of occultations covers transits, eclipses and occultations of one object by the other in the same planetary system. I will describe each of the classes of occultations and what we can learn from them.

Occultation of a Star

The amount of starlight drops abruptly when an airless body occults a star. In this case, the starlight is not bent. See Figure 6.18. The situation is different though when a planet like Uranus blocks out sunlight. In this case, the planet's atmosphere bends (or refracts) the star light. See Figure 6.19. Very thin layers of gas can cause a large amount of refraction. For Uranus and Neptune, gases at the 1 to 10 μbar level can cause significant refraction. The amount of refraction depends on temperature, chemical composition, density and the wavelength of light used. Therefore, one can obtain information on these quantities form occultation data.

In cases where the center of the planet moves in front of the star's image, we may see a central flash. The central flash is a brightening that occurs in the middle of an occultation. The positions of the Earth and the occulting body needed for a central flash are shown in Figure 6.20. The atmosphere near the edges of the occulting body refracts extra star light directly opposite from the star, and this is what causes the central flash. Astronomers observed central flashes for three Neptune occultations, and a Nov. 4, 1998, Triton occultation.

One can use occultation and central flash data to learn more about the temperature, density, pressure, chemical composition and transparency of another planet's atmosphere. Occultation measurements taken over several years can also

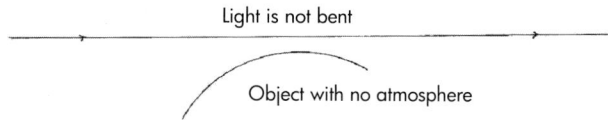

Figure 6.18. When a body with no atmosphere moves near the path of starlight, the light continues to move in a straight direction as shown. The intensity of the starlight drops abruptly as the airless body blocks the path of light. (Credit: Richard W. Schmude, Jr.)

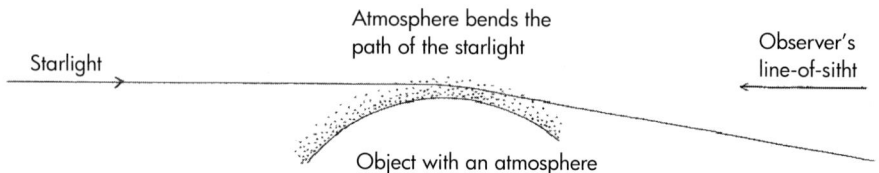

Figure 6.19. When a planetary atmosphere gets close to the path of starlight, it will refract or bend the light causing it to change direction. The starlight will gradually dim as seen from the Earth as thicker parts of the atmosphere bend starlight more and more. (Credit: Richard W. Schmude, Jr.)

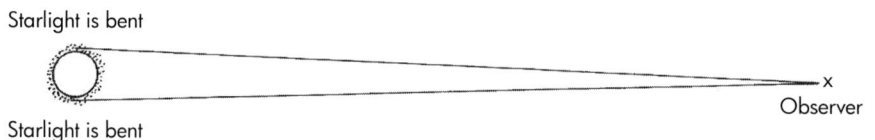

Figure 6.20. When the observer moves directly behind a planetary atmosphere that is blocking starlight, a central flash will be seen. This is because many parts of the planetary atmosphere bend light towards the observer. (Credit: Richard W. Schmude, Jr.)

show how these quantities change over time. One also can use occultation data to search for very thin atmospheres around Charon and the larger moons of Uranus. Finally, when people record occultation data at different locations, they can determine both the shape of the occulting body and its exact position.

I will describe a few successful occultation experiments that amateur astronomers carried out from three different continents.

Pluto moved in front of a faint star (magnitude 15.5) on June 12, 2006. Several teams of astronomers in Australia and New Zealand successfully recorded the occultation. One group of professional astronomers measured the occultation with the 3.9 meter Anglo-Australia telescope (AAT). They used the AAT data to determine the thickness of Pluto's atmosphere. Other astronomers at several locations obtained additional occultation data which were used to constrain the shadow path of the occultation. This additional data were especially important for the operators of the AAT telescope. Two amateur astronomers from Australia, who participated in the occultation study, were Dave Gault and Blair Lade. I will discuss how Dave and Blair obtained their data.

Dave Gault used a 0.25 m (10 inch) Newtonian telescope along with a Meade Deep Sky Imager Pro camera to record his data. On May 14, 2006, Dave produced an image of Pluto under nearly a full moon. He made this image to simulate the conditions on the night of the Pluto occultation, which would be during a nearly full moon one month later. Later in the month, Dave planned to image Pluto on June 12 in the hope of detecting an occultation by one of its small moons. The preliminary prediction was that Pluto's shadow would miss Dave's location, but, as it turned out, Pluto's path crossed his location. He measured the occultation and his data were used in determining the path of Pluto's shadow. The lesson here is that occultation predictions have some uncertainty, and even if the occultation path is predicted to miss, it may be worthwhile to record data in case the prediction is wrong.

Blair Lade spent two weeks planning for the June 12, 2006 Pluto occultation. He used a 0.5 m (20 inch) f/5 Newtonian telescope along with a Meade DSI Pro black and white camera to record the occultation. See Figure 6.21. The telescope belongs to the Astronomical Society of South Australia. He purchased several accessories and also made trial runs. Blair also used the software package "Tardis" to get the correct time from a GPS onto a computer as there was no internet connection at the observatory. The time the images were taken is embedded in the FITS header by the DSI camera's application. During the occultation, Blair had his computer take a 1.0 second image of Pluto every other second for about 2.5 hours. The images were automatically stored on a hard disc. Since each image required 1.2 megabytes and approximately 5000 images were made, over six gigabytes of memory was required. He also took 100 bias frames, 100 dark frames and 100 flat frames at the end of the observing session. The images were stored in FITS format. He analyzed the images and constructed a graph of flux versus time. There was an obvious drop in flux as Pluto blocked out the light from the star. More importantly, the drop in flux was gradual, which is consistent with Pluto having an atmosphere.

Occultation measurements can also yield important negative data, such as the lack of rings around a planet or the lack of an atmosphere. I would like to present two examples of important negative data.

Antonio Cidadao recorded the occultation of the star HIP 106829 by Titania on Sept. 8, 2001, from Portugal. He used a 0.25 m Schmidt-Cassegrain telescope along with an SBIG ST-237 CCD camera. He recorded 9999 images between

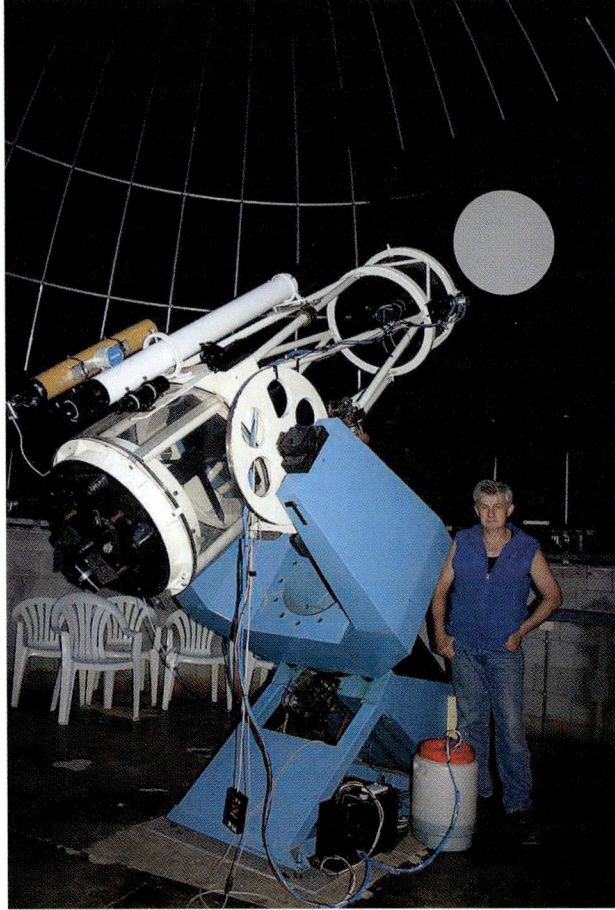

Figure 6.21. A picture of Blair Lade and the 0.5 meter (20 inch) telescope which he used to measure Pluto's occultation of a star on June 12, 2006. (Credit: Blair Lade.)

1:37:39 UT and 3:27:46 UT. Each image had an exposure time of 0.1 seconds and was corrected for both bias and dark current. Images were made every 0.301 seconds. The brightness of the star was measured on images recorded near the occultation time. The resulting light curve is shown in Figure 6.22. The drop in intensity occurred within 0.6 seconds instead of 20 to 30 seconds for Pluto. An upper limit of 1.0 microbar was selected for the surface pressure of Titania's atmosphere.

The writer also made measurements of HIP 106829 on Sep. 8, 2001. I used an SSP-3 photometer along with a filter transformed to the Johnson V system and a 0.5 m Newtonian telescope. Uranus, Titania and HIP 106829 were placed into the photometer FOV and brightness measurements were made every ten seconds. All measurements were made at Villa Rica, Georgia, in the United States. I was looking for any opaque objects near the orbits of Titania and Oberon, such as rings. An opaque object would have caused the star to dim, resulting in a drop in the photometer reading. No large object occulted the star from my location.

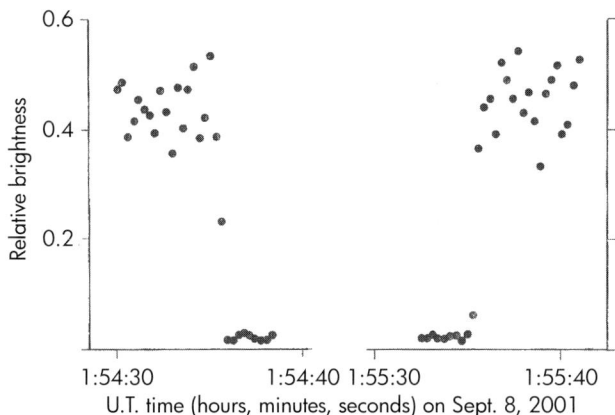

Figure 6.22. A light curve of the star HIP 106829 as it was occulted by Titania on Sept. 8, 2001. The light curve was constructed by the writer from data recorded by Antonia Cidadao. (Credit: Antonio Cidadao.)

Occultations by Earth's Moon

One can obtain useful data of Uranus and Neptune when the moon occults these objects. Richard Radick and William Tetley, for example, used a high speed photoelectric photometer to record the brightness of Uranus in 1977 as it reappeared on the dark side of the Moon. They used a filter with a peak transmission of 690 nm and with a FWHT of 45 nm to record their data. Scattered light from the Moon was a problem, but it did not stop them from detecting limb darkening and polar brightening on Uranus. They also measured a radius of 25,700 ± 500 km for that planet, which is close to the accepted value.

Transits, Eclipses, and Occultations

Rarely have astronomers observed transits, eclipses and occultations of moons, rings and planets within the Uranus and Neptune system. As a result, we will focus more on what we can learn from these events. Similar events for other planetary systems are also discussed in the hope that some will study transits, eclipses, and occultations in the Uranus, Neptune, and Pluto systems.

At rare times, the moons of Uranus, Neptune or Pluto can move in front of one another, cast shadows on one another, transit the planet that they orbit or move into the planet's shadow. These events can be seen only when the observer passes very close to the orbital plane of the affected moon. In the case of Uranus, the five largest moons have orbits which are nearly in the planet's equatorial plane; hence, when the observer passes near the equatorial plane of Uranus, he/she will be able to observe these events. The Earth will pass through Uranus's equatorial plane in Dec. 2007 and Feb. 2050, and mutual events will occur near these times. The situation for Triton, Charon and many of the smaller moons is more complicated, because they do not move in the equatorial plane of the planets that they orbit. The last set of mutual events between Pluto and Charon took place during the 1980s. These events gave astronomers information about the size, density, composition and color of the two objects.

Table 6.10. The different kinds of satellite transits, occultations, and eclipses that can occur under the right conditions

Description	Figure number
Satellite transits a second satellite	6.23A
Satellite shadow (umbra portion) transits a second satellite	6.23B
Satellite shadow (penumbral portion) transits a second satellite	6.23C
Satellite moves completely in front of a second satellite	6.23D
Satellite moves in front of just part of a second satellite	6.23E
Satellite moves in front of rings	6.23F
Satellite moves behind the rings	6.23 G
Satellite moves in front of a planet	6.23 H
Satellite shadow moves in front of a planet	6.23 I
Satellite moves into the planet's (penumbral portion) shadow	6.23 J
Satellite moves into the planet's (umbra portion) shadow	6.23 K
Satellite moves behind the planet that it orbits	6.23 L

Table 6.10 lists the different kinds of satellite events that can occur, and Figure 6.23 illustrates these events. All of the events in Table 6.10 will create a drop in brightness that depends on the satellite characteristics, the wavelength of light used and the geometrical alignment of the Sun, observer and bodies involved in the event.

Apostolos Christou carried out a thorough study of the orbits of Uranus's five largest moons with the goal of predicting satellite mutual events. His results were published in late 2005. He lists the times and predicted magnitude drops for the most visible 78 events between 2006 and 2009. Measurements of these events will enable astronomers to determine the positions of the moons to within 0.02 arc-seconds. Astronomers can use this position data to compute the gravitational perturbations which these moons exert on each other. Astronomers with large telescopes will be able also to obtain information about large albedo features in the northern hemispheres of these moons. One group of astronomers measured three occultations of Europa by Io, and used the data to draw up a map of Europa years before Voyager imaged that moon.

Tony Mallama was able to determine the altitude of Jupiter's north polar haze by studying a partial eclipse of Callisto near Jupiter's northern limb. The same type of event can be used to obtain information about the haze layers on Uranus and Neptune.

How can one observe the satellite mutual events and transits of the moons of Uranus and Neptune? First, an accurate time signal is essential. The most important piece of information that one obtains in an eclipse or transit is the time of the event. There are several ways of obtaining the accurate time. One way is to call WWV at 303-499-7111 and listen to the time signals on the phone. One can also use a Global Positioning Satellite (GPS) instrument to obtain the time. Finally, one can use software like "Tardis" to feed the accurate time into the computer. Secondly, one should use a CCD camera since it is easier to subtract scattered light from the planet. Unlike Jupiter's bright moons, those of Uranus are much fainter. Longer exposure times and larger apertures are required for Uranus's faint moons. According to Tony Mallama, three additional things must be done during an eclipse measurement, namely, try to get a second uneclipsed moon in each

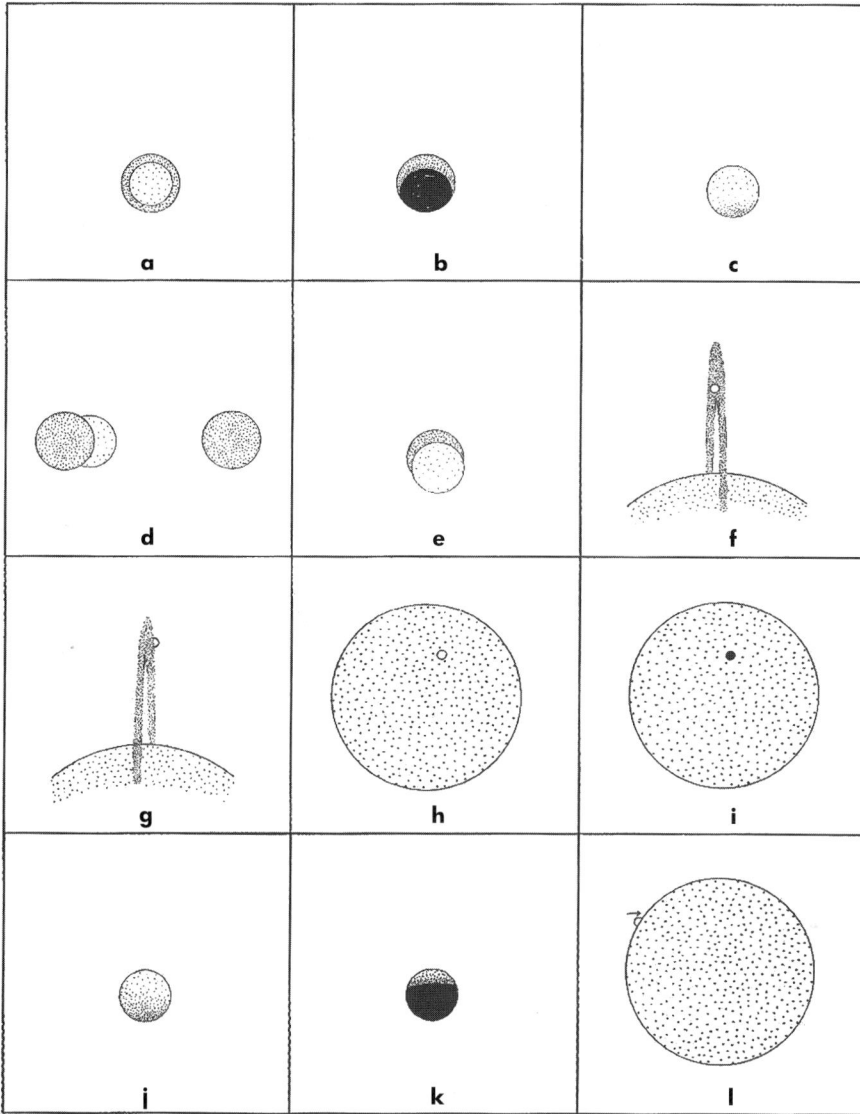

Figure 6.23. Different kinds of occultations and eclipses that one can study: Frame A, A smaller and more reflective moon moves in front of a second moon; Frame B, the shadow of a smaller moon covers part of a second moon; Frame C, the penumbral portion of one moon's shadow covers a second moon; Frame D, a larger and less reflective moon blocks out a second moon; Frame E, A smaller and more reflective moon covers part of a larger moon, Frame F, a moon moves in front of a planet's rings; Frame G, a moon moves behind a planet's rings; Frame H, a moon moves in front of the planet that it orbits; Frame I, the shadow of a moon transits a planet; Frame J, a moon moves into a planet's shadow (penumbral portion) Frame K, a moon moves into a planet's shadow (umbra portion) Frame L, a moon is eclipsed by the planet it orbits. (Credit: Richard W. Schmude, Jr.)

frame which contains the eclipsed moon, use a filter and use a telescope that lets through as little scattered light as possible. A refractor, a closed-tube Newtonian or a Schmidt-Cassegrain are all excellent choices for eclipse studies.

One should experiment with different filters and exposure times before making measurements of an eclipse of a faint moon. The objective of an eclipse measurement is to measure the time of minimum brightness. I would start with a wideband filter that has a transmission centered at 900 nm in the near-infrared. Uranus and Neptune are much dimmer at a wavelength of 900 nm compared to visible light, but yet the moons are brighter at infrared wavelengths. The exposure times should be as short as possible; however, they should show a measureable signal for the moon. There is a delicate balance between short exposure and a measureable signal for which the observer should strive. One can start with a 30 second exposure with a 0.2 meter (8 inch) telescope and see what kind of moon image shows up. The time may be shorter for larger telescopes.

Satellite Brightness Measurements

Uranus's four largest moons – Ariel, Umbriel, Titania and Oberon – can brighten by over a factor of two in green light between conjunction and opposition. Most of this change is due to their large opposition surges. These moons are discussed in Chapter 1. Table 6.11 shows the brightness of Uranus's s four largest moons near opposition in green light. As an example, Titania shines at magnitude 13.7 on opposition night but it drops to magnitude 13.9 five days later. The values in Table 6.11 apply to the year 2010, but can be used up to 2020 with little error provided that the albedos of the northern hemispheres of these moons are similar to those of their southern hemispheres. The large moons of Uranus also display large brightness changes in near-infrared light. Therefore one can measure the brightness changes of these moons in different colors of light. The best kind of measurement is made with a standard filter like one transformed to the Johnson V system.

Three ALPO members, Charles Bell, Ed Grafton and Frank Melillo, have successfully measured the relative brightness of two or more moons of Uranus. All three made their measurements from unfiltered, digital images.

Table 6.11. Brightness of the four brightest moons of Uranus at different times near opposition for the years 2008–2020

Days Before/After Opposition	Solar Phase Angle (deg.)	Ariel	Umbriel	Titania	Oberon
0	<0.1	13.9	14.7	13.7	13.9
1	<0.1	14.0	14.7	13.7	13.9
2	0.1	14.1	14.8	13.8	14.0
3	0.2	14.1	14.8	13.8	14.0
4	0.2	14.2	14.9	13.9	14.1
5	0.2	14.2	14.9	13.9	14.1
10	0.5	14.3	15.0	14.0	14.2
20	1.0	14.4	15.0	14.1	14.3
40	2.0	14.5	15.2	14.3	14.5
90	2.9	14.6	15.3	14.4	14.6

All four of Uranus's large moons undergo synchronous rotation. As a result of this, the same side of each of these faces Uranus and, hence, different hemispheres face the Earth. One can measure the brightness of these moons as they move around their primary. The best times to do this would be between 2008 and 2013 and between 2044 and 2054 when the equatorial regions of these moons face Earth.

Appendix

Measuring Transformation Coefficients

There are two ways of measuring transformation coefficients; these are the Two-Star Method and the Multiple-Star Method. The Two-Star Method is easy to use, but the Multiple-Star Method is more accurate. One should make transformation measurements under excellent sky conditions and at a temperature near that of what measurements are made. Transformation measurements should be made near 10°C (50°F).

Multiple-Star Method

In this method, one must measure the brightness of several stars with different B–V values. The stars should be as close to one another as possible, and be as close to the observer's zenith as possible. This will insure that extinction corrections will be small. The selected stars should not change in brightness by more than ∼0.02 magnitudes. (It is the writer's opinion that most of the stars visible to us, including the Sun, undergo at least small changes in brightness from time to time.) Once measurements are made, one must compute values of ΔV, Δv and $k' \times \Delta AM$ for each of the target stars in the same way as described in Chapter 5. The heart of this method is Equation A.1:

$$\varepsilon_V \times \Delta(B - V) = \Delta V - \Delta v + k' \times \Delta AM + k'' \times AM_{avg} \times \Delta(B - V) \qquad (A.1)$$

where ε_V is the transformation coefficient, AM is the air mass and other letters and symbols are defined in the same way as in Chapter 5. (This is the same as equation 13.10.6 in Hall and Genet, ©1988, p. 200.) The $k'' \times AM_{avg} \times \Delta(B-V)$ term is negligible except for the U and B filters. In order to transform a filter to the Johnson V system, one should plot average values of $\Delta V - \Delta v + k' \times \Delta AM$ versus the values of $\Delta(B-V)$ and compute the slope (ε_V) or, if the data is not linear, one should use a curve fitting routine to develop an equation which will yield a value of ε_V

Here is how one carries out measurements using the Multiple-Star Method with an example. As a first step, I selected the constellation Ursa Major because it was near my zenith. Afterwards, I checked to see if stars in this constellation changed in brightness by a large amount using the Millennium Star Atlas and I avoided any stars listed in the Atlas as variables. After this, I selected several bright stars in Ursa Major. On May 9, 2007, under clear skies, I carried out V filter measurements of

Table A.1 Stars used in my multiple-star analysis made on May 9, 2007

Star	V Magnitude [a]	B–V [a]
Eta-Ursae Majoris (η-UMa)	1.86	−0.18
Delta-Ursae Majoris (δ-UMa)	3.33	0.08
Gamma-Ursae Majoris (γ-UMa)	2.44	0.00
Beta-Ursae Majoris (β-UMa)	2.38	−0.01
Alpha-Ursae Majoris (α-UMa)	1.80	1.07
Theta-Ursae Majoris (θ-UMa)	3.20	0.46
Iota-Ursae Majoris (ι-UMa)	3.15	0.19

[a] Magnitude and B–V values are from Iriarte et al (1965).

them with the goal of measuring the transformation coefficient of my system. Eta (η) Ursae Majoris, one of the seven stars, was the comparison star in this study. The magnitudes and B−V color indexes of the selected stars are listed in Table A.1. I used the method described in Chapter 5 to measure values of ΔV, Δv and $k' \times \Delta AM$ for the other six stars ($k' = 0.2419$ magnitudes/air mass). The results determined on May 9, 2007, are shown in Table A.2.

After computing the ΔV, Δv and $k' \times \Delta AM$ values, I computed average values of these quantities and used them to compute average values of $\Delta V - \Delta v + k' \times \Delta AM$ for each of the six stars. Average values of Δv, $k' \times \Delta AM$ and $\Delta V - \Delta v + k' \times \Delta AM$ are listed in Table A.2. Afterwards I constructed a graph of $\Delta V - \Delta v + k' \times \Delta AM$ versus $\Delta(B-V)$ and determined the best fit curve. As it turned out, I was able to fit the data to a linear equation using a least squares approach. The resulting equation was:

$$\Delta V - \Delta v + k' \times \Delta AM = 0.0145 - 0.061 \times \Delta(B - V) \tag{A.2}$$

and the slope, ε_V, was equal to -0.061.

Two-Star Method

In order to measure transformation coefficients with the Two-Star Method, one should pick out two stars near each other with different B−V values. These stars

Table A.2 A summary of measurements made on May 9, 2007, for the purpose of determining a transformation coefficient using the Multiple-Star Method

Star	ΔV	Average (Δv)	Average ($k' \times \Delta AM$)	$\Delta V - \Delta v + k' \times \Delta AM$	$\Delta(B-V)$
(η-UMa)	0.00	0.0	0.0	0.00	0.00
(δ-UMa)	1.47	1.467	0.005	0.008	0.26
(γ-UMa)	0.58	0.579	0.007	0.008	0.18
(β-UMa)	0.52	0.525	0.017	0.012	0.17
(α-UMa)	−0.06	0.026	0.023	−0.063	1.25
(θ-UMa)	1.34	1.413	0.046	−0.027	0.64
(ι-UMa)	1.29	1.378	0.077	−0.011	0.37

should be as close to the zenith as possible and should have a nearly constant brightness. One can rearrange equation A.1 as:

$$\varepsilon_V = (\Delta V - \Delta v + k' \times \Delta AM)/\Delta(B - V) \qquad (A.3)$$

where the $k'' \times AM_{avg} \times \Delta(B - V)$ term is dropped since it is negligible for the V filter. If Eta-Ursae Majoris and Alpha-Ursae Majoris are used, $\Delta V = -0.06 - 0.00 = -0.06$; $\Delta v = 0.026 - 0.00 = 0.026$; $k' \times \Delta AM = 0.023$; $\Delta B - V = 1.07 - -0.18 = 1.25$; and the resulting transformation coefficient, based on the appropriate average values would be:

$$\varepsilon_V = (-0.06 - 0.026 + 0.023)/1.25 = -0.050 \qquad (A.4.)$$

One problem with the Two-Star Method is that it implies a linear relation between the values of $\Delta V - \Delta v + k' \times \Delta AM$ and $\Delta(B-V)$ terms. However, this may not be the case. As a rule of thumb, if the transformation coefficient exceeds 0.1, one should use the Multiple-Star Method to measure transformation coefficients. If one desires an accuracy of 0.002 magnitudes or better, he or she should use the Multiple-Star Method.

Measuring Extinction Coefficients

The extinction coefficient shows how much light is absorbed by the atmosphere per air mass. Its units are magnitude/air mass. The extinction coefficient is different for different wavelengths of light. Between 2004 and 2007, I measured average extinction coefficient values of 0.38, 0.23, 0.16 and 0.12 magnitude/air mass for filters transformed to the Johnson B, V, R and I system. All measurements were made under clear skies in central Georgia.

Does the extinction coefficient change from one night to the next? Yes! For example, I measured extinction coefficients of 0.184, 0.242 and 0.218 magnitude/air mass for three excellent nights on May 8, 9 and 10, 2007. All measurements were made through a filter that was transformed to the Johnson V system. I believe that extinction coefficients should be measured each time that magnitude measurements are made unless the difference in air mass between the target and comparison star is less than 0.1 air masses. If the target and comparison star are very close to each other, extinction corrections would be negligible.

There are two ways of measuring the extinction coefficient which I call the Drift Method and the Two-Object Method.

Drift Method

Let us describe this method with an example. On May 10, 2007, I measured the brightness of Venus as it was setting. The data are summarized in Table A.3. I computed the altitude (A) of Venus from the equation:

$$\text{Inverse} \sin(A) = \cos(\theta) \times \cos(\phi) \times \cos(h) + \sin(\theta) \times \sin(\phi) \qquad (A.5)$$

In this equation, θ is the observer's latitude (33.1°N in my case), ϕ is the declination of the target (26.0° for Venus on May 10, 2007) and h is the hour angle which is how far (in degrees) the target is from the observer's meridian. The

Table A.3 Venus data collected with the purpose of determining the extinction coefficient using the Drift Method

Time (U.T.)	Delta Time (minutes)	h (degrees)	Altitude (degrees)	Air Mass (AM) [a]	Diff.	Δmag
1:04	268	67	32.25	1.874	2566.33	0.000
1:13	277	69.25	30.41	1.976	2503.33	0.027
1:26	290	72.5	27.76	2.147	2337.33	0.101
1:54	318	79.5	22.12	2.655	2134.33	0.200
2:06	330	82.5	19.74	2.961	2066.67	0.235
2:19	343	85.75	17.17	3.388	1901.33	0.326
2:30	354	88.5	15.02	3.859	1768.33	0.404
2:41	365	91.25	12.88	4.485	1492.0	0.589
2:48	372	93	11.54	5.000	1332.67	0.711

[a] One can compute the Air Mass value for Venus from: Air Mass = 1/Sin(A) where A is the altitude of Venus.

Astronomical Almanac lists when each of the planets transit the meridian for people located at longitude of $0°$ on Earth. On May 10, 2007, Venus transited the meridian at 14:59 U.T. for people at $0°$ longitude. My longitude at the time of the measurements was $84.14°W$ and, since Earth rotates 15 degrees/hour, Venus transited my meridian at: 14:59 U.T. + ($84.14°/15°$ per hour) = 14:59 U.T. + (5.609 hours) = 20:36 U.T. (Recall that 5.609 hours is approximately equal to 5 hours and 37 minutes.) I computed the hour angle, h, using:

$$h(\text{in degrees}) = (\text{delta time in minutes})/4 \qquad (A.6)$$

where delta time is the difference in time between the meridian transit, 20:36 U.T., and the time of measurement. Values for h and delta time are listed in Table A.3.

I computed the Diff. values of Venus for each set of measurements using:

$$\text{Diff} = (\text{average Venus reading}) - (\text{average sky reading}) \qquad (A.7)$$

Afterwards, I computed the Δmag values. The Δmag value is the difference in brightness between the first Venus reading at 1:04 and the other readings. For example, Δmag for the second reading at 1:13 U.T. is:

$$\Delta\text{mag} = 2.5\text{xlog}(2566.33/2503.33) = 0.027\text{magnitudes}$$

The other Δmag values are computed in the same way and are listed in Table A.3.

Finally I determined the extinction coefficient by plotting the Δmag values versus the air mass values and used a linear least squares routine to obtain:

$$\Delta\text{mag} = -0.398\text{magnitudes} + (0.218\text{magnitude/airmass}) \times \text{airmass} \qquad (A.8)$$

The slope of this line, 0.218 magnitude/air mass, is the extinction coefficient.

Two-Object Method

A second and quicker method of measuring the extinction coefficient is to use the Two-Object Method. One simply measures the brightness of two objects of known

Name: _____

Address: _____

e-mail: _____

Telescope type: _____   Magnification: _____

Seeing: _____   Transparency: _____

Drawing

*Lable N for north sky direction and P
for preceding edge of the disc.*

◯   ◯

Drawing   Intensities

Planet: _____

Date/Time (U.T.): _____
Write comments blow

Color Estimate

Planet: _____

Color: _____

Date: _____

Time (U.T.): _____
Write comments blow.

Visual magnitude estimates

Comparison star magnitudes: _____

Source of comparison star magnitudes: _____

Planet: _____   Estimated magnitude: _____

Instrument used: _____

Date/Time (U.T.): _____

Fig. A.1 The official observation form of the Remote Planets Section of ALPO.

brightness at different altitudes and then measures the magnitude difference, which is due to extinction. These objects can be either stars or planets.

Let's describe this method with an example. Table A.4 lists the data of two objects that I measured on May 8, 2007. I used the same procedure in Chapter 5 to compute Diff. and Δv. I let Beta-Ophiuchi be the comparison star in this example.

Table A.4 Measurements and analysis of two stars made in order to determine the extinction coefficient using the Two-Object Method

Star	Time U.T.	B–V	Diff.	ΔV	Δv	Air Mass [a]
Beta-Ophiuchi	8:36	1.16	54.33	–	–	1.141
Theta-Scorpii	9:01	0.41	74.83	–0.91	–0.348	4.408

[a] One can use the equation Air Mass = 1/sin(A) where A is the altitude to compute the Air Mass value of the star. The only difference is that for stars, one looks for the closest planet to the star and looks up when that planet crosses the central meridian. Therefore, one adds (or subtracts) the difference in right ascension between the star and the closest planet on the date of the measurement in order to compute the time when the star transits the meridian at 0° longitude.

I rearranged Equation A.1 as:

$$k' = (\varepsilon_V \times \Delta(B - V) - \Delta V + \Delta v)/\Delta AM \tag{A.9.}$$

(Keep in mind that the $k'' \times AM_{avg} \times \Delta(B - V)$ term drops out for the V filter.) I computed $\Delta(B-V)$ as: $\Delta(B-V)$ = the B−V value of Theta-Scorpii minus the B−V value of Beta-Ophiuchi or $0.41 - 1.16 = -0.75$. We know that $\Delta v = 2.5 \log [54.33/74.83] = -0.348$, and the ΔAM term equals $4.408 - 1.141 = 3.267$ air masses. After substitution:

$$k' = [(-0.051 \times -0.75) - -0.91 + -0.348]\text{magnitudes}/(3.267\text{air masses}) \tag{A.10}$$

or: $k' = 0.184$ magnitude/air mass. In this calculation, I used a value of $\varepsilon_V = -0.051$.

The Two-Object Method is much quicker and is the one that I use usually. The measurements should be done at nearly the same time that magnitude measurements are made since the sky transparency can change during the night.

ALPO Observation Form

Figure A.1 shows the official observation form of the Remote Planets Section of the Association of Lunar and Planetary Observers (ALPO). The form may be reproduced for personal use only. When making an observation, image or measurement, it is important to send as much relevant information as practicable. It is especially important for the observer to describe the location of the north direction of his/her sky and the location of the preceding limb somewhere near any drawing.

Bibliography

Agnor CB and Hamilton DP (2006) 'Neptune's Capture of its Moon Triton in a Binary-Planet Gravitational Encounter,' Nature 441: 192–194.

Alexander AFO'D (1965) The Planet Uranus, American Elsevier Pub. Co. Inc., New York.

Astronomical Almanac, U.S. Govt. Printing Office, Washington DC.

Baines KH, and Hammel HB (1994) 'Clouds, Hazes, and the Stratospheric Methane Abundance in Neptune,' Icarus 109: 20–39.

Baines KH, Mickelson ME, Larson LE et al (1995) 'The Abundances of Methane and Ortho/Para Hydrogen on Uranus and Neptune: Implications of New Laboratory 4-0 H_2 Quadrupole Line Parameters,' Icarus 114: 328–340.

Baron RL, French RG, Elliot JL (1989) 'The Oblateness of Uranus at the 1-μbar Level,' Icarus 78: 119–130.

Bauer JM, Roush TL, Geballe TR et al (2002) 'The Near Infrared Spectrum of Miranda: Evidence of Crystalline Water Ice,' Icarus, 158: 178–190.

Beatty JK (1988) 'Discovering Pluto's Atmosphere,' Sky and Telescope, 76 (6): 624–627.

Beatty JK and Chaikin A – editors (1990) The New Solar System, 3rd edn. Cambridge University Press, Cambridge.

Behannon KW, Lepping RP, Sittler EC et al (1987) 'The Magnetotail of Uranus,' J. Geophys. Res. 92: 15,354–15,366.

Belton MJS, Wallace L, Hayes SH et al (1980) 'Neptune's Rotation Period: A Correction and a Speculation on the Difference between Photometric and Spectroscopic Results,' Icarus 42: 71–78.

Benton JL Jr (2005) Saturn and How to Observe it, Springer-Verlag, London.

Bergstralh JT, Miner ED, Matthews MS - editors (1991) Uranus, The University of Arizona Press, Tucson.

Bézard B, Romani PN, Feuchtgruber H et al (1999) 'Detection of the Methyl Radical on Neptune,' The Astrophysical Journal 515: 868–872.

Binzel RP, Mulholland JD (1984) 'Photometry of Pluto During the 1983 Opposition: A New Determination of the Phase Coefficient,' The Astronomical Journal 89: 1759–1761.

Bishop R (2003) 'Binoculars,' in Observer's Handbook 2004, The University of Toronto Press, Toronto, pp. 52–55.

Bockelée-Morvan D, Lellouch E, Biver N et al (2001) 'Search for CO gas in Pluto, Centaurs and Kuiper Belt objects at radio wavelengths,' Astronomy and Astrophysics 377: 343–353.

Bowen KP (1985) 'Binocular Astronomy: Is there a Difference?' Sky and Telescope 69: 572–573.

Brown ME, Koresko D, Blake GA (1998) 'Detection of Water Ice on Nereid,' The Astrophysical Journal 508: L175–L176.

Brown RH (1983) 'The Uranian Satellites and Hyperion: New Spectrophotometry and Compositional Implications,' Icarus 56: 414–425.

Brown RH, and Cruikshank DP (1983) 'The Uranian Satellites: Surface Compositions and Opposition Brightness Surges,' Icarus 55: 83–92.

Brown RH, Johnson TV, Goguen JD et al (1991) 'Triton's Global Heat Budget,' Science 251: 1465–1467.

Brown TL, LeMay HE Jr, Bursten BE et al (2006) Chemistry: The Central Science, 10th edn. Prentice Hall, Upper Saddle River, NJ.

Budine PW (1988) 'Jupiter's Oscillating Spot of 1987,' in Proceedings of the Astronomical League and the A. L. P. O. (Van Zandt RP-editor) pp. 13–15.

Buie MW, Grundy WM (2000) 'The Distribution and Physical State of H_2O on Charon,' Icarus 148: 324–339.

Buie MW, Grundy WM, Young EF et al (2006) 'Orbits and Photometry of Pluto's Satellites: Charon, S/2005 P1 and S/2005 P2,' The Astronomical Journal 132: 290–298.

Buie MW, Tholen DJ, Wasserman LH (1997) 'Separate Lightcurves of Pluto and Charon,' Icarus 125: 233–244.

Buratti BJ, Goguen JD, Mosher JA (1997) 'No Large Brightness Variations on Nereid' Icarus 126: 225–228.

Buratti BJ, Hillier JK, Heinze A et al (2003) 'Photometry of Pluto in the Last Decade and Before: Evidence for Volatile Transport?' Icarus 162: 171–182.

Burgdorf M, Orton GS, Davis GR et al (2003) 'Neptune's Far-Infrared Spectrum form the ISO Long-Wavelength and Short-Wavelength Spectrometers,' Icarus 164: 244–253.

Burgdorf M, Orton G, van Cleve, J et al (2006) 'Detection of New Hydrocarbons in Uranus' Atmosphere by Infrared Spectroscopy,' Icarus 184: 634–637.

Charnoz S, Deau E, Brahic A et al (2005) 'Saturn's F ring seen by CASSINI ISS: Several Strands or a Single Spiral?' Bulletin of the American Astronomical Society 37: 768.

Chavez CE, Murray CD, Beurle K et al (2005) 'Saturn's F Ring: The Role of Prometheus in the Production of Regular azimuthal Structure,' Bulletin of the American Astronomical Society 37: 768.

Cain L (1984) 'A 17 ½-inch Binocular Reflector,' Sky and Telescope 68: 460–463.

Christou AA (2005) 'Mutual Events of the Uranian Satellites 2006–2010,' Icarus 178: 171–178.

Cook JC, Desch SJ, Roush TL et al (2006) 'Near-Infrared Spectroscopy of Charon: Possible Evidence for Cryovolcanism on Kuiper Belt Objects,' Bulletin of the American Astronomical Society 38: 518.

Cook JC, Desch SJ, Roush TL et al (2007) 'Near-Infrared Spectroscopy of Charon: Possible Evidence for Cryovolcanism on Kuiper Belt Objects,' The Astronomical Journal 663: 1406–1419.

Countin R, Gautier D, Strobel D (1996) 'The CO Abundance on Neptune from HST Observations,' Icarus 123: 37–55.

Cross EW Jr (1969) 'Some Visual Observations of Markings on Uranus,' J.A.L.P.O. 21 (9–10): 152–153.

Cruikshank DP, Matthews MS and Schumann AM (1995) Neptune and Triton, The University of Arizona Press, Tucson.

Cruikshank D, Mason RE, Dalle Ore CM et al (2006) 'Ethane on Pluto and Triton,' Bulletin of the American Astronomical Society 38: 518.

Cruikshank D, Schmitt B, Roush TL et al (2000) 'Water Ice on Triton,' Icarus 147: 309–316.

DeBoer DR, Steffes PG (1994) 'Laboratory Measurements of the Microwave Properties of H_2S under Simulated Jovian Conditions with an Application to Neptune,' Icarus 109: 352–366.

DeBoer DR, Steffes PG (1996) 'Estimates of the Tropospheric Vertical Structure of Neptune Based on Microwave Radiative Transfer Studies,' Icarus 123: 324–335.

de Pater I, Gibbard SG, Hammel HB (2006) 'Evolution of the Dusty Rings of Uranus,' Icarus, 180: 186–200.

de Pater I, Gibbard SG, Macintosh BA et al (2002) 'Keck Adaptive Optics Images of Uranus and its Rings,' Icarus 160: 359–374.

de Pater I, Hammel H, Showalter MR et al (2007) 'First results from Ground-based Observations of the Ring Plane Crossings of Uranus,' Bulletin of the American Astronomical Society 39: 427.

Delitsky M (2006) 'Triton's Ionosphere: Chemistry and Composition,' Bulletin of the American Astronomical Society 38: 519.

Dobbins TA, Parker DC, Capen CF (1988) Observing and Photographing the Solar System, Willmann-Bell, Inc., Richmond.

Duxbury NS, Brown RH (1997) 'The Role of an Internal Heat Source for the Eruptive Plumes on Triton,' Icarus 125: 83–93.

Elliot JL (1979) 'Stellar Occultation Studies of the Solar System,' Annual Review of Astronomy and Astrophysics, 17: 445–475.

Elliot JL, Ates A, Babcock BA et al (2003) 'The recent expansion of Pluto's atmosphere,' Nature 424: 165–168.

Elliot JL, Dunham E (1977) 'Discovering the Rings of Uranus,' Sky and Telescope 53 (6): 412–416 and 430.

Elliot JL, Nicholson PD (1984) 'The Rings of Uranus,' in Planetary Rings (Greenberg, R. and Brahic, A. – Editors) The University of Arizona Press, Tucson, pp. 25–72.

Elliot JL, Person MJ, Gulbis AA et al (2006) 'The Size of Pluto's Atmosphere As Revealed by the 2006 June 12 Occultation,' Bulletin of the American Astronomical Society 38: 541.

Elliot JL, Person MJ, Gulbis AA et al (2007) "Changes in Pluto's Atmosphere: 1988–2006,' The Astronomical Journal 134: 1–13.

Elliot JL, Strobel DF, Zhu, X et al (2000) 'The Thermal Structure of Triton's Middle Atmosphere,' Icarus 143: 425–428.

Encrenaz T, Kallenbach R, Owen TC et al (2005) 'The Outer Planets and Their Moons, Springer, Dordrecht, The Netherlands.

Encrenaz T, Schulz B, Drossart P et al (2000) 'The ISO Spectra of Uranus and Neptune Between 2.5 and 4.2 μm: Constraints on Albedos and H_3^+,' Astronomy and Astrophysics 358: L83–L87.

Ferrari C, Brahic A (1994) 'Azimuthal Brightness Asymmetries in Planetary Rings,' Icarus 111: 193–210.

Feuchtgruber H, Lellouch E, Bézard B et al (1999) 'Detection of HD in the Atmospheres of Uranus and Neptune: A New Determination of the D/H Ratio,' Astronomy and Astrophysics 341: L17–L21.

Fix JD (2006) Astronomy: Journey to the Cosmic Frontier, 4th edn, McGraw Hill, Boston.

Flynn B, Stern SA, Trafton L et al (2001) 'HST/FOC Imaging of Triton,' Icarus 150: 297–302.

Fouchet T, Lellouch E, Feuchtgruber H (2003) 'The Hydrogen Ortho-to-Para Ratio in the Stratospheres of the Giant Planets,' Icarus: 161: 127–143.

French RG, Elliot JL, Frenach LM et al (1988) 'Uranian Ring Orbits from Earth-Based and Voyager Occultation Observations,' Icarus 73: 349–378.

French RG, Elliot JL, Sicardy B et al (1982) 'The Upper Atmosphere of Uranus: A Critical Test of Isotropic Turbulence Models,' Icarus 51: 491–508.

French RG, McGhee CA, Sicardy B (1998) 'Neptune's Stratospheric Winds from Three Central Flash Occultations,' Icarus 136: 27–49.

French RG, Melroy PA, Baron RL et al (1985) 'The 1983 June 15 Occultation by Neptune II. The Oblateness of Neptune,' The Astronomical Journal 90: 2624–2638.

French RG, Roques F, Nicholson PD et al (1996) 'Earth-Based Detection of Uranus' Lambda Ring,' Icarus 119: 269–284.

Fry PM, Sromovsky LA (2007) 'Uranus Cloud Layers As Constrained By HST STIS Spectra,' Bulletin of the American Astronomical Society 39: 526.

George D (1995) 'Starting Out Right,' CCD Astronomy, pp. 18–23, Summer 1995.

Gibbard SG Roe H, de Pater I et al (2002) 'High-Resolution Infrared Imaging of Neptune from the Keck Telescope,' Icarus 156: 1–15.

Gibbard SG, de Pater I, Roe HG et al (2003) 'The Altitude of Neptune Cloud Features from High-Spatial-Resolution Near-Infrared Spectra,' Icarus 166: 359–374.

Goguen JD, Hammel HB, Brown RH (1989) 'V Photometry of Titania, Oberon and Triton,' Icarus 77: 239–247.

Goldreich P, Porco CC (1987) 'Shepherding of the Uranian Rings. II. Dynamics,' The Astronomical Journal 93: 730–737.

Grav T, Holman MJ, Fraser WC (2004) 'Photometry of Irregular Satellites of Uranus and Neptune,' The Astrophysical Journal 613: L77–L80.

Gray HJ, Isaacs A (1975) A New Dictionary of Physics, Longman Group Ltd., London.

Grundy WM, Buie MW (2001) 'Distribution and Evolution of CH_4, N_2 and CO Ices on Pluto's Surface: 1995 to 1998,' Icarus 153: 248–263.

Grundy WM, Buie MW (2002) 'Spatial and Compositional Constraints on Non-ice Components and H_2O on Pluto's Surface,' Icarus 157: 128–138.

Grundy WM, Buie MW, Spencer JR (2002) 'Spectroscopy of Pluto and Triton at 3–4 Microns: Possible Evidence for Wide Distribution of Nonvolatile Solids,' The Astronomical Journal 124: 2273–2278.

Grundy WM, Noll KS, Stephens DC (2005) 'Diverse Albedos of Small Trans-Neptunian Objects,' Icarus 176: 184–191.

Grundy WM, Young LA (2004) 'Near-Infrared Spectral Monitoring of Triton with IRTF/SPeX I: Establishing a Baseline for Rotational Variability,' Icarus 172: 455–465.

Grundy WM, Young LA, Spencer JR et al (2006) 'Distributions of H_2O and CO_2 ices on Ariel, Umbriel, Titania and Oberon from IRTF/SpeX Observations,' Icarus 184: 543–555.

Grundy WM, Young LA, Young EF (2003) 'Discovery of CO_2 Ice and Leading-Trailing Spectral Asymmetry on the Uranian Satellite Ariel,' Icarus, 162: 222–229.

Guggenheim EA (1967) Thermodynamics, North-Holland, Amsterdam.

Gulbis AA, Elliot JL, Person MJ et al (2006) 'Pluto's Atmospheric Structure: Results From The 2006 June 12 Stellar Occultation,' BAAS 38: 541.

Gulbis AA, Elliot JL, Person MJ et al (2006) 'Charon's Radius and Atmospheric Constraints from Observations of a Stellar Occultation,' Nature 439: 48–51.

Gurwell MA, Butler BJ (2005) 'Sub-Arcsecond Scale Imaging of the Pluto/Charon Binary System at 1.4 mm,' Bulletin of the American Astronomical Society 37: 743.

Haitchuck RH, Henden AA, Truax R (1994) 'Photometry in the Digital Age,' CCD Astronomy, pp. 20–23, Fall 1994.

Hall DS, Genet RM (1988) Photoelectric Photometry of Variable Stars, 2nd edn, Willmann-Bell Inc., Richmond.

Hammel HB, de Pater I, Gibbard SG et al (2005a) 'New Cloud Activity on Uranus in 2004: First Detection of a Southern Feature at 2.2 μm,' Icarus 175: 284–288.

Hammel HB, de Pater I, Gibbard SG et al (2005b) 'Uranus in 2003: Zonal Winds, Banded Structure and Discrete Features,' Icarus 175: 534–545.

Hammel HB, Lark NL, Rigler M et al (1989) 'Disk-Integrated Photometry of Neptune at Methane-Band and Continuum Wavelengths,' Icarus 79: 1–14.

Hammel HB, Lawson SL, Harrington J et al (1992) 'An Atmospheric Outburst on Neptune from 1986 through 1989,' Icarus 99: 363–367.

Hammel HB, Lockwood GW (1997) 'Atmospheric Structure of Neptune in 1994, 1995 and 1996: HST Imaging at Multiple Wavelengths,' Icarus 129: 466–481.

Hammel HB, Lockwood GW (2007) 'Long-Term Atmospheric Variability on Uranus and Neptune,' Icarus 186: 291–301.

Hammel HB, Lockwood GW, Mills JR et al (1995) 'Hubble Space Telescope Imaging of Neptune's Cloud Structure in 1994,' Science 268: 1740–1742.

Hammel HB, Sitko ML, Orton GS et al (2007) 'Infrared Imaging of Neptune with Gemini/Michelle and Keck/NIRC2,' Bulletin of the American Astronomical Society 39: 527.

Hampel CA, Hawley GG - Editors, (1973) The Encyclopedia of Chemistry, 3rd edn, Van Nostrand Reinhold Company, New York.

Harrington P (2001) 'Everything You Need to Know About Binoculars,' Astronomy 29 (6): 68–77.

Harris DL (1961) 'Photometry and Colorimetry of Planets and Satellites,' in Planets and Satellites (Kuiper GP, Middlehurst BM-editors) The University of Chicago, Chicago, pp. 272–342.

Hawkins RL (1996) 'Super System! Assembling an Ideal CCD System for Photometry,' CCD Astronomy, pp. 14–17, Spring 1996.

Helfenstein P, Thomas PC, Veverka J (1989) 'Evidence from Voyager II Photometry for Early Resurfacing of Umbriel,' Nature, 338: 324–326.

Henden AA, Kaitchuck RH (1990) 'Astronomical Photometry,' 2nd edn, Willmann-Bell, Inc., Richmond.

Hester J, Burstein D, Blumenthal G et al (2007) 21st Century Astronomy, 2nd edn, W.W. Norton & Co., Inc., New York.

Hicks MD, Buratti BJ (2004) 'The Spectral Variability of Triton from 1997–2000,' Icarus 171: 210–218.

Hill RE – editor (1990) The New Observe and Understand the Sun, The Astronomical League, Washington, DC.

Hillier J, Helfenstein P, Veverka J (1989) 'Miranda: Color and Albedo Variations from Voyager Photometry,' Icarus, 82: 314–335.

Hillier J, Veverka J (1994) 'Photometric Properties of Triton Hazes,' Icarus 109: 284–295.

Hillier J, Veverka J, Helfenstein P et al (1994) 'Photometric Diversity of Terrains on Triton,' Icarus 109: 296–312.

Hirshfeld A, Sinnott RW, Ochsenbein F (1991) Sky Catalogue 2000.0 Volume 1: Stars to Magnitude 8.0, Sky Publishing Corp., Cambridge, MA.

Hollis AJ (1989) 'Uranus 1954 to 1986,' The Journal of the British Astronomical Association 99: 59–62.

Holman MJ, Kavelaars JJ, Grav T et al (2004) 'Discovery of five Irregular Moons of Neptune,' Nature 430: 865–867.

Holmes A (1995) 'Optimizing a CCD Imaging System,' CCD Astronomy, p. 14–18, Winter 1995.

Hopkins J (1976) Glossary of Astronomy and Astrophysics, The University of Chicago Press, Chicago.

Horn LJ, Yanamandra-Fisher PA, Esposito LW et al (1988) 'Physical Properties of Uranian Delta Ring from a Possible Density Wave,' Icarus 76: 485–492.

Hubbard WB, Nellis WJ, Mitchell AC et al (1991) 'Interior Structure of Neptune: Comparison with Uranus,' Science 253: 648–651.

Hubbard WB, Nicholson PD, Lellouch E et al (1987) 'Oblateness, Radius, and Mean Stratospheric Temperature of Neptune from the 1985 August 20 Occultation,' Icarus 72: 635–646.

Hunter T (1999) 'What Can Go Wrong: Observatory Mistakes to Avoid,' Sky and Telescope 97: 132.

IAU Circular No. 8676 (2006).

IAU Circular No. 8686 (2006).

IAU Circular No. 8802 (2007).

IAU Circular No. 8826 (2007).

Iriarte B, Johnson HL, Mitchell RI et al (1965) 'Five-Color Photometry of Bright Stars,' Sky and Telescope, 30 (No. 1): 21–31.

Jacobson RA, Owen WM Jr (2004) 'The Orbits of the Inner Neptunian Satellites from Voyager, Earth-Based and Hubble Space Telescope Observations,' The Astronomical Journal 128: 1412–1417.

Janesick J, Blouke M (1987) 'Sky on a Chip: The Fabulous CCD,' Sky and Telescope, 74 (3): 238–242.

Janesick J (1997) 'CCDs: The Inside Story,' CCD Astronomy, pp. 10–15, Winter 1997.

Jensen D (1976) The Principles of Physiology, (Illustrated by B. Jensen) Appleton-Century-Crofts, New York.

Johnson HL, Mitchell RI, Iriarte B et al (1966) 'No. 63 UBVRIJKL Photometry of the Bright Stars,' Communications of the Lunar and Photometry Laboratory. Vol. 4, No. 3.

Joos F, Schmid HM (2007) 'Limb Polarization of Uranus and Neptune II. Spectropolarimetric Observations,' Astronomy and Astrophysics 463: 1201–1210.

Joyce RR, Pilcher CB, Cruikshank DP et al (1977) 'Evidence for Weather on Neptune I.' The Astrophysical Journal 214: 657–662.

Karkoschka E (1997) 'Rings and Satellites of Uranus: Colorful and Not So Dark,' Icarus, 125: 348–363.

Karkoschka E (1999) 'Small Satellites of Uranus: Not So Small, Except the New One,' Bulletin Of the American Astronomical Society 31:1074.

Karkoschka E (2001a) 'Comprehensive Photometry of the Rings and 16 Satellites of Uranus with the Hubble Space Telescope,' Icarus 151: 51–68.

Karkoschka E (2001b) 'Voyager's Eleventh Discovery of a Satellite of Uranus and Photometry and the First Size Measurements of Nine Satellites,' Icarus 151: 69–77.

Karkoschka E (2003) 'Sizes Shapes, and Albedos of the Inner Satellites of Neptune,' Icarus 162: 400–407.

Kavelaars JJ, Holman MJ, Grav T et al (2004) 'The Discovery of Faint Irregular Satellites of Uranus' Icarus 169: 474–481.

Kelly P (2006) 'Observer's Handbook 2007,' Toronto ON: The Royal Astronomical Society of Canada.

Klein MJ, Hofstadter MD (2006) 'Long-Term Variations in the Microwave Brightness Temperature of the Uranus Atmosphere,' Icarus 184: 170–180.

Lee MH, Peale SJ (2006) 'On the Orbits and Masses of the Satellites of the Pluto-Charon System,' Icarus 184: 573–583.

Lellouch E, Paubert G, Moreno R et al (2000) 'Search for Variations in Pluto's Millimeter-Wave Emission,' Icarus 147: 580–584.

Lellouch E, Romani PN, Rosenqvist J (1994) 'The Vertical Distribution and Origin of HCN in Neptune's Atmosphere,' Icarus 108: 112–136.

Lellouch E, Stansberry J, Cruikshank D et al (2006) 'Pluto's Thermal Lightcurve: Spitzer Mips and Irs Observations,' Bulletin of the American Astronomical Society 38: 518.

Li J, Zhou LY, Sun YS (2007) 'The Origin of the High-Inclination Neptune Trojan 2005 TN$_{53}$,' Astronomy and Astrophysics 464: 775–778.

Lindal GF, Lyons JR, Sweetnam DN et al (1987) 'The Atmosphere of Uranus: Results of Radio Occultation Measurements with Voyager 2,' Journal of Geophysical Research. 92: 14,987–15,001.

Livitski R (1993) 'How I Built a 20-inch Binocular,' Sky and Telescope 85: 89–91.

Lockwood GW (1977) 'Secular Brightness Increases of Titan, Uranus, and Neptune 1972–1976,' Icarus 32: 413–430.

Lockwood GW, Jerzykiewicz M (2006) 'Photometric Variability of Uranus and Neptune, 1950–2004,' Icarus 180: 442–452.

Lockwood GW, Thompson DT (1999) 'Photometric Variability of Uranus, 1972–1996,' Icarus 137: 2–12.

Lockwood GW, Thompson DT (2002) 'Photometric Variability of Neptune, 1972–2000,' Icarus 156: 37–51.

Lunine JI (1993) 'The Atmospheres of Uranus and Neptune,' Annual Review of Astronomy and Astrophysics. 31: 217–263.

MacRobert AM (1992) 'A Pupil Primer,' Sky and Telescope 83: 502–504.

MacRobert AM (1983) 'Backyard Astronomy – 2 Observing with Binoculars,' Sky and Telescope 66: 309–310.

Mallama A (1992) 'Observing Jupiter Satellite Eclipses with a CCD Camera,' Sky and Telescope 83: 700–703.

Mallama A, Krobusek B, Collins DF et al (2002), 'Jupiter: CCD Photometry of Galilean Satellites,' J.A.L.P.O. 44 (2): 15–21.

Marchi S, Barbieri C, Lazzarin M et al (2004), 'A 0.4–2.5 μm Spectroscopic Investigation of Triton's Two Faces,' Icarus 168: 367–373.

Marieb EN (1995) Human Anatomy and Physiology, The Benjamin/Cummings Publishing Company, Inc., Redwood City, CA.

Maris M, Carraro G, Cremonese G et al (2001) 'Multicolor Photometry of the Uranus Irregular Satellites Sycorax and Caliban,' The Astronomical Journal 121: 2800–2803.

Marten A, Matthews HE, Owen T et al (2005) 'Improved Constraints on Neptune's Atmosphere from Submillimetre-Wavelength Observations,' Astronomy and Astrophysics 429, 1097–1105.

Martin S, de Pater I, Kloosterman J ct al (2006) 'Multi Wavelength Imaging of Neptune at High Spatial Resolution,' Bulletin of the American Astronomical Society 38: 502.

Max CE, Macintosh BA, Gibbard SG et al (2003) 'Cloud Structures on Neptune Observed with Keck Telescope Adaptive Optics,' The Astronomical Journal 125: 364–375.

Mayer EH (1984) 'Finder Follies,' Sky and Telescope 67: 210.

McEwen AS (1990) 'Global Color and Albedo Variations on Triton,' Geophysical Research Letters 17: 1765–1768.

Meadows V, Orton G, Liang MC et al (2006) 'First Spitzer Observations of Neptune: Detection of New Hydrocarbons,' Bulletin of the American Astronomical Society 38: 502.

Meeus J (1997) 'Equinoxes and Solstices on Uranus and Neptune,' Journal of the British Astronomical Association 107: 332.

Meinke BK, Esposito LW, Coldwell, JE (2006) 'Moonlets and Clumps in Saturn's F Ring,' Bulletin of the American Astronomical Society 38: 572.

Michalsky JJ, Stokes RA (1977) 'Whole-Disk Polarization Measurements of Uranus at Visible Wavelengths,' The Astrophysical Journal 213: L135–L137.

Michels WC, Hoyt RC, May JC et al (1961) The International Dictionary of Physics and Electronics, Van Nostrand Company Inc., New York.

Miner ED (1998) Uranus: The Planet, Rings and Satellites, 2nd edn, Praxis Publishing, Chichester, UK.

Miner ED, Wessen RR (2002) Neptune: The Planet, Rings and Satellites, Praxis Publishing Ltd., Chichester UK.

Minton RB (1972) 'Latitude Measures of Jupiter in the 0.89μ Methane Band,' Communications of the Lunar and Planetary Laboratory 9:339–352 (Comm. No. 176).

Moses JI, Naphas RD, Greathouse TK (2005) 'Time Variation of Hydrocarbon Abundances on Uranus and Neptune: Predictions from a 1-D Model,' Bulletin of the American Astronomical Society 37: 680.

Moses JI, Rages K, Pollack JB (1995) 'An Analysis of Neptune's Stratospheric Haze Using High-Phase-Angle Voyager images,' Icarus 113: 232–266.

Murray CD, Evans MW, Cooper N et al (2005) 'Saturn's F Ring and its Retinue,' Bulletin of the American Astronomical Society 37: 767–768.

Mutchler MJ, Stern SA, Weaver HA et al (2006) 'The B-V Colors and Photometric Variability of Nix and Hydra, Pluto's Two Small Satellites,' Bulletin of the American Astronomical Society 38: 542.

Nagler A (1991) 'Choosing Your Telescope's Magnification,' Sky and Telescope 81: 553–559.

Nesvorný D, Dones L (2002) 'How Long-Lived Are the Hypothetical Trojan Populations of Saturn, Uranus and Neptune?' Icarus 160: 271–288.

Newberry MV (1994a) 'The Signal to Noise Connection,' CCD Astronomy, pp. 34–39, Summer 1994.

Newberry MV (1994b) 'The Signal to Noise Connection Part II,' CCD Astronomy, pp. 12–15, Fall 1994.

Newberry MV (1995a) 'Recovering the Signal,' CCD Astronomy, pp. 18–21, Spring 1995.

Newberry MV (1995b) 'Dark Frames,' CCD Astronomy, pp. 12–14, Summer 1995.

Newberry MV (1996) 'Pursuing the Ideal Flat Field,' CCD Astronomy, pp. 18–21, Winter, 1996.

Nicholson PD (2006) 'Natural Satellites of the Planets,' in Observer's Handbook 2007, Toronto: The University of Toronto Press, pp. 20–26.

Nicholson PD, Mosqueira I (1995) 'Stellar Occultation Observations of Neptune's Rings: 1984–1988,' Icarus 113: 295–330.

North G (2004) Observing Variable Stars, Novae and Supernovae, Cambridge University Press, Cambridge.

Norton OR (2002) The Cambridge Encyclopedia of Meteorites, Cambridge University Press, Cambridge.

Norwood J, Chanover NJ (2007) 'Latitudinal Variations in Uranus' Near –Infrared Methane Absorption Bands,' Bulletin of the American Astronomical Society 39: 526.

Oberc P (1994) 'Dust Impacts Detected by Voyager 2 at Saturn and Uranus: A Post-Halley View,' Icarus 111: 211–226.

O'Meara SJ (2000) 'Seeking the Elusive Gegenschein,' Sky and Telescope 100 (4): 116–119.

Olkin CB, Elliot JL, Hammel HB et al (1997) 'The Thermal Structure of Triton's Atmosphere: Results from the 1993 and 1995 Occultations,' Icarus 129, 218–201.

Olkin CB, Wasserman LH, Franz OG (2003) 'The Mass Ratio of Charon to Pluto from Hubble Space Telescope Astrometry with the Fine Guidance Sensors,' Icarus 164: 254–259.

Olkin CB, Young EF, Young LA et al (2005) 'Evidence of Tholins on Pluto's Surface from Near-IR and IR Spectroscopy,' Bulletin of the American Astronomical Society 37: 743.

Olkin CB, Young EF, Young LA et al (2007) 'Pluto's Spectrum from 1.0 to 4.2 μm: Implications for Surface Properties,' The Astronomical Journal 133: 420–431.

Optec, Inc. (1997) Model SSP-3 Solid-State Stellar Photometer Technical Manual for Theory of Operation and Operating Procedures, Revision three, Optec. Inc., Lowell, MI.

Orton G, Burgdorf M, Meadows V et al (2005) 'First Results of Middle-Infrared Spectroscopy of Uranus and Neptune from Spitzer,' Bulletin of the American Astronomical Society 37: 662.

Parker DC, Dobbins TA (1987a) 'The Art of Planetary Observing – I,' Sky and Telescope 74: 370–372.

Parker DC, Dobbins TA (1987b), 'The Art of Planetary Observing – II,' Sky and Telescope, 74: 603–607.

Pascu D, Storrs AD, Wells EN et al (2006) 'HST BVI Photometry of Triton and Proteus,' Icarus 185: 487–491.

Pascu D, Rohde JR, Seidelmann PK et al (1998) 'Hubble Space Telescope Astrometric Observations and Orbital Mean Motion Corrections for the Inner Uranian Satellites,' The Astronomical Journal 115: 1190–1194.

Pascu D, Rohde JR, Seidelmann PK (1998) 'HST BVI Photometry of Triton and Proteus,' Bulletin of the American Astronomical Society 30: 1101.

Peale SJ (1999) 'Origin and Evolution of the Natural Satellites,' Annual Review of Astronomy and Astrophysics, 37: 533–602.

Peek BM (1981) The Planet Jupiter, revised edn, Faber and Faber Ltd., London.

Person MJ, Elliot JL, Gulbis AA et al (2007) 'High Altitude Structure in Pluto's Atmosphere from the 2007 March 18 Stellar Occultation,' Bulletin of the American Astronomical Society 39: 519–520.

Person MJ, Elliot JL, Gulbis AA et al (2006) 'Charon's Radius and Density from the Combined Data Sets of the 2005 July 11 Occultation,' The Astronomical Journal 132: 1575–1580.

Pesnell WD, Grebowsky JM, Weisman AL (2004) 'Watching Meteors on Triton,' Icarus 169: 482–491.

Plescia JB (1987) 'Cratering History of the Uranian Satellites: Umbriel, Titania and Oberon,' Journal of Geophysical Research, 92: 14,918–14,932.

Pocock G, Richards CD (2004) Human Physiology the Basis of Medicine, 2nd edn, Oxford University Press, Oxford.

Podolak M, Reynolds RT, Young R (1990) 'Post Voyager Comparisons of the Interiors of Uranus and Neptune,' Geophysical Research Letters 17: 1737–1740.

Porco CC, Cuzzi JN, Ockert ME et al (1987) 'The Color of the Uranian rings,' Icarus 72: 69–78.

Porco CC Goldreich P (1987) 'Shepherding of the Uranian Rings. I. Kinematics,' Astron. J. 93: 724–729.

Protopapa S, Boehnhardt H, Herbst T et al (2007) 'Surface Ice Spectroscopy of Pluto and Charon Resolved,' Bulletin of the American Astronomical Society 39: 541–542.

Rages KA, Hammel HB, Friedson AJ (2004) 'Evidence for Temporal Change at Uranus' South Pole,' Icarus 172: 548–554.

Rages K, Hammel HB, Lockwood GW (2002) 'A Prominent Apparition of Neptune's South Polar Feature,' Icarus 159: 262–265.

Rages K, Pollack JB (1992) 'Voyager Imaging of Triton's Clouds and Hazes,' Icarus 99: 289–301.

Rages K, Pollack JB, Tomasko MG et al (1991) 'Properties of Scatterers in the Troposphere and Lower Stratosphere of Uranus Based on Voyager Imaging Data,' Icarus 89: 359–376.

Reeves R (2006) Introduction to Webcam Astrophotography, Willmann-Bell, Inc., Richmond.

Reitsema HJ, Hubbard WB, Lebofsky LA et al (1982) 'Occultation by a Possible Third Satellite of Neptune,' Science 215: 289–291.

Roddier F, Roddier C, Graves JE et al (1998) 'Neptune's Cloud Structure and Activity: Ground-Based Monitoring with Adaptive Optics,' Icarus 136: 168–172.

Roe HG, Graham JR, McLean IS et al (2001) 'The Altitude of an Infrared-Bright Cloud Feature on Neptune from Near-Infrared Spectroscopy,' The Astronomical Journal 122: 1023–1029.

Rogers JH (1995) 'The Giant Planet Jupiter,' Cambridge University Press, Cambridge.

Romon J, de Bergh C, Barucci MA et al (2001) 'Photometric and Spectroscopic Observations of Sycorax, Satellite of Uranus,' Astronomy and Astrophysics 376: 310–315.

Ruhland CT, Blow G, Broughton J et al (2006) 'The Pluto stellar Occultation of 2006 June 12: Observations and Joint Analysis,' Bulletin of the American Astronomical Society 38: 541.

Ruiz J (2003) 'Heat Flow and Depth to a Possible Internal Ocean on Triton,' Icarus 166: 436–439.

Schaefer BE, and Schaefer, MW (2000) 'Nereid Has Complex Large-Amplitude Photometric Variability,' Icarus 146: 541–555.

Schmid HM, Joos F, Tschan D (2006) 'Limb polarization of Uranus and Neptune I. Imaging Polarimetry and Comparison with Analytic Models,' Astronomy and Astrophysics 452: 657–668.

Schmude RW Jr (1996) 'Uranus and Neptune in 1993,' Journal of the Association of Lunar and Planetary Observers 39 (2): 63–66.

Schmude RW Jr (1998) 'Seasonal Changes in Sporadic Meteor Rates,' Icarus 135: 496–500.

Schmude RW Jr (2001) 'Observations of the Remote Planets in 2000,' Journal of the Association of Lunar and Planetary Observers 43 (3): 30–37.

Schmude RW Jr (2002a) 'The 1991–92 Apparition of Jupiter,' Journal of the Association of Lunar and Planetary Observers 44 (2): 22–40.

Schmude RW Jr (2002b) 'The Uranus, Neptune and Pluto Apparitions in 2001,' Journal of the Association of Lunar and Planetary Observers 44 (3): 22–31.

Schmude RW Jr (2003) 'Jupiter: A Report on the 2001–2002 Apparition,' Journal of the Association of Lunar and Planetary Observers 45 (2): 41–62.

Schmude RW Jr (2004) 'The Uranus, Neptune and Pluto Apparitions in 2002,' Journal of the Association of Lunar and Planetary Observers 46 (4): 47–55.

Schmude RW Jr (2005) 'The Uranus, Neptune and Pluto Apparitions in 2003,' Journal of the Association of Lunar and Planetary Observers 47 (2): 38–43.

Schmude RW Jr (2006) 'Uranus, Neptune and Pluto: Observations During the 2004 Apparitions,' Journal of the Association of Lunar and Planetary Observers 48 (2): 41–45.

Schmude RW Jr (2006) 'The Remote Planets in 2005–06,' Journal of the Association of Lunar and Planetary Observers 48 (4): 32–37.

Schmude R Jr, Melillo F (2006) 'Observing Uranus and Its Moons,' Sky and Telescope 112 (4): 75–78.

Schulz B, Encrenaz T, Bézard B et al (1999) 'Detection of C_2H_4 in Neptune from ISO/PHT-S observations,' Astronomy and Astrophysics 350: L13–L17.

Seeley RR, Stephens TD, Tate P (2006) Anatomy and Physiology, 7th edn, McGraw Hill Higher Education, Boston.

Shartle SM (1968) 'The Ellipticity of Uranus,' Journal of the Association of Lunar and Planetary Observers 20 (11–12): 197–200.

Sheehan W, O'Meara SJ (1993) 'Exotic Worlds,' Sky and Telescope 85: 20–24.

Showalter MR, Lissauer JJ, de Pater I (2005) 'The Rings of Neptune and Uranus in the Hubble Space Telescope,' Bulletin of the American Astronomical Society 37: 772.

Showalter MR, Lissauer JJ, de Pater I (2006) 'Outer Rings and Chaotic Orbits in the Uranian System,' Bulletin of the American Astronomical Society 38: 553.

Showalter MR, Lissauer JJ, French RG et al (2007) 'HST Observations of the Uranian Ring Plane Crossing: Early Results,' Bulletin of the American Astronomical Society 39: 427.

Sicardy B, Bellucci A, Gendron E et al (2006) 'Charon's Size and an Upper Limit on its Atmosphere from a Stellar Occultation,' Nature 439: 52–54.

Sicardy B, Combes M, Brahic A et al (1982) 'The 15 August 1980 Occultation by The Uranian System: Structure of the Rings and Temperature of the Upper Atmosphere,' Icarus 52: 454–472.

Sicardy B, Widemann T, Lellouch E et al (2003) 'Large Changes in Pluto's Atmosphere as Revealed by Recent Stellar Occultations,' Nature 424: 168–170.

Sinnott RW, Perryman MAC (1997) Millennium Star Atlas, Sky Publishing Corporation, Cambridge.

Sky and Telescope 99: 96 'The Near Sky: Problems with Airglow,' 2000.

Sky and Telescope 105: 27 'Neptune's Growing Family' 2003.

Smirnov OM, Ipatov AP, Samus NN (1995) 'PSF Photometry,' CCD Astronomy, Pages 14–17, Spring, 1995.

Smith BA, Smith SA (1972) 'Upper Limits for an Atmosphere on Io,' Icarus 17: 218–222.

Smith FG, King TA (2000) Optics and Photonics: An Introduction, John-Wiley and Sons Ltd., Chichester.

Smith MD, Gierasch PJ (1995) 'Convection in the Outer Planet Atmospheres Including Ortho-Para Hydrogen Conversion,' Icarus 116: 159–179.

Soderblom LA, Kieffer SW, Becker TL et al (1990) 'Triton's Geyser-Like Plumes: Discovery and Basic Characterization,' Science 250: 410–415.

Sromovsky LA (2005) 'Accurate and Approximate Calculations of Raman Scattering in the Atmosphere of Neptune,' Icarus 173: 254–283.

Sromovsky LA, Fry PM (2005) 'Dynamics of Cloud Features on Uranus,' Icarus 179: 459–484.

Sromovsky LA, Fry PM, Baines KH (2002) 'The Unusual Dynamics of Northern Dark Spots on Neptune,' Icarus 156: 16–36.

Sromovsky LA, Fry PM, Baines KH et al (2001) 'Coordinated 1996 HST and IRTF Imaging of Neptune and Triton II. Implications of Disk-Integrated Photometry,' Icarus 149: 435–458.

Sromovsky LA, Fry PM, Dowling TE et al (2001a) 'Coordinated 1996 HST and IRTF Imaging of Neptune and Triton III. Neptune's Atmospheric Circulation and Cloud Structure,' Icarus 149: 459–488.

Sromovsky LA, Fry PM, Dowling TE et al (2001b) 'Neptune's Atmospheric Circulation and Cloud Morphology: Changes Revealed by 1998 HST Imaging,' Icarus 150: 244–260.

Sromovsky LA, Fry PM, Limaye SS et al (2003) 'The Nature of Neptune's Increasing Brightness: Evidence for a Seasonal Response,' Icarus 163: 256–261.

Sromovsky LA, Limaye SS, Fry PM (1995) 'Clouds and Circulation on Neptune: Implications of 1991 HST Observations,' Icarus 118: 25–38.

Stanton RH (1999) 'Visual Magnitudes and the "Average Observer": The SS-Cygni Field Experiment,' Journal of the A.A.V.S.O. 27 (2): 97–112.

Steffey PC (1982) 'Red Stars and Staring,' Sky and Telescope 64: 213.

Steffey PC (1992) 'The Truth About Star Colors,' Sky and Telescope 84: 266–273.

Steffl AJ, Stern SA (2007) 'First Constraints on Rings in the Pluto System,' The Astronomical Journal 133: 1485–1489.

Sterken C, Manfroid J (1992) Astronomical Photometry: A Guide, Kluwer Academic Publishers, Dordrecht, The Netherlands.

Stern A, Mitton J (1998) Pluto and Charon, John Wiley and Sons, Inc., New York.

Stern SA, Tholen DJ-editors (1997) Pluto and Charon, The University of Arizona Press, Tucson.

Stevens SS (1961) 'To Honor Fechner and Repeal His Law,' Science 133: 80–86.

Stone EC, Cooper JF, Cummings AC et al (1986) 'Energetic Charged Particles in the Uranian Magnetosphere,' Science 233: 93–97.

Strom RG, Croft SK, Boyce JM (1990) "The Impact Cratering Record on Triton,' Science 250: 437–439.

Tholen DJ, Buie MW, Grundy WM (2007) 'Dynamical State of the Pluto System,' Bulletin of the American Astronomical Society 39: 542.

Thomas PC (2000) 'The Shape of Triton from Limb Profiles,' Icarus 148: 587–588.

Thomas P, Veverka J, Johnson TV et al (1987) 'Voyager Observations of 1985U1,' Icarus, 72: 79–83.

Trafton LM, Miller S (2004) 'Images of Uranus' H_3^+ Emission,' Bulletin of the American Astronomical Society 36: 1072–1073.

Trilling DE, Brown RH (2000) 'Red Gray and Blue: Near Infrared Spectroscopy of Faint Moons of Uranus and Neptune,' Icarus 148: 301–306.

Vashkov'yak MA (1999) 'Evolution of the Orbits of Distant Satellites of Uranus,' Astronomy Letters 25: 476–481 (English Translation).

Vermilion JR, Clark RN, Greene TF et al (1974) 'A Low Resolution Map of Europa from four Occultations by Io,' Icarus 23: 89–96.

Veverka J, Thomas P, Helfenstein P et al (1987) 'Satellites of Uranus: Disk-Integrated Photometry from Voyager Imaging Observations,' Journal of Geophysical Research 92: 14,895–14,904.

Vilas F, Lederer SM, Gill SL et al (2006) 'Aqueous Alteration Affecting the Irregular Outer Planets Satellites: Evidence from Spectral Reflectance,' Icarus 180: 453–463.

Wasserman L, Veverka J (1973) 'On the Reduction of Occultation Light Curves,' Icarus 20: 322–345.

Winter Giuliatti, SM, Ferreira Gonsalves MA, Winter O et al (2005) 'Comments on the Interaction Between the F Ring System and its New Discovered Objects,' Bulletin of the American Astronomical Society 37: 768.

Woodhead JA (2001) Earth Science Volume I: The Physics and Chemistry of Earth, Pasadena CA: Salem Press Inc.

Xiaolong D, LeBeau RP (2007) 'Dynamic Simulations of Potential Dark Spots in the Atmosphere of Uranus,' Bulletin of the American Astronomical Society 39: 525–526.

Young AT (1990) 'How We Perceive Star Brightness,' Sky and Telescope 79: 311–313.

Young EF, Buie MW, French RG et al (2006) 'The Detailed Vertical Structure of Pluto's Atmosphere from the 12 Jun 2006 Stellar Occultation,' Bulletin of the American Astronomical Society 38: 541–542.

Young EF, Buie MW, Young LA (2005) 'Results from PIXON-Processed HRC images of Pluto,' Bulletin of the American Astronomical Society 37: 743.

Young EF, Young LA, Buie M (2007) 'Pluto's Radius,' Bulletin of the American Astronomical Society 39: 541.

Young L, Buie M, French R et al (2006) 'Physical Processes in Pluto's Atmosphere from its 2006 June 12 Occultation,' Bulletin of the American Astronomical Society 38: 542.

Young LA, French RG, Gregory B et al (2005) 'New Occultation Systems and the 2005 July 11 Charon Occultation,' Bulletin of the American Astronomical Society 37: 743–744.

Zeilik M (1997) Astronomy: The Evolving Universe, 8th edn, John Wiley and Sons Inc., New York.

Index

Index

Printed in the United States